A Biologist's Basic Mathematics

David R. Causton
M.Sc., Ph.D., D.I.C.

Department of Botany & Microbiology,
University College of Wales, Aberystwyth

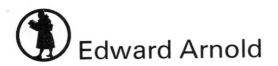
Edward Arnold

First published in Great Britain 1983
by Edward Arnold (Publishers) Ltd
41 Bedford Square
London WC1B 3DQ

Edward Arnold (Australia) Pty Ltd
80 Waverley Road
Caulfield East 3145
PO Box 234
Melbourne

First Published in United States of America 1983
By Edward Arnold
300 North Charles Street
Baltimore
Maryland 21201

British Library Cataloguing in Publication Data

Causton, David R.
  A biologist's basic mathematics.
  (A series of student texts in contemporary biology).
  Vol. 1
  1. Biomathematics
  I. Title   II. Series
  510'.24574   QH323.5

ISBN 0-7131-2879-8

Text set in Photon Times
by Spottiswoode Ballantyne Ltd., Colchester and London

Preface

This book is the first of two which are effectively the 'second edition' of my *A Biologist's Mathematics*, first published in 1977. Having regard to the needs of biology students with respect to mathematical training and knowledge, it is now considered desirable to have, on the one hand, a text containing only the basic mathematical principles and methods that are mostly required by the large majority of first and second year students, and on the other, a more advanced book required by some final year undergraduates, postgraduates, and research workers. The present book corresponds to the more elementary two-thirds of *A Biologist's Mathematics*, while the sequel – *A Biologist's Advanced Mathematics* is a largely new work, but incorporating the more advanced sections of the former book.

After an introductory chapter on the role of mathematics in biology, the book considers the fundamental properties of numbers, indices, and logarithms, in Chapter 2. Chapter 3 is also a foundation chapter, dealing with the basis of the geometrical interpretation of many aspects of mathematics relevant to the biologist. A full multi-dimensional approach is adopted; and the topics of location in space, the measurement of distances between points, and linear functions and their geometric representation are covered, examples being given of the biological use of each topic. Chapter 4 then considers non-linear functions and their curves, including the Michaelis–Menten function and allometric relationships, and also contains a section on some general properties of curves to serve as an introduction to the following chapters on calculus.

The next three chapters deal with the calculus. Chapter 5 describes principles and methods of the differential calculus, while Chapter 6 is concerned with physical interpretation and usage. Chapter 7 deals with the two aspects of integration: the indefinite integral and simple methods of integration, and the definite integral.

Hitherto, the principal type of mathematical function used is the polynomial, although other kinds of function of potential interest to the biologist are introduced in Chapter 4. Exponential and related functions, which are of outstanding importance to biologist and mathematician alike are described in Chapter 9. Although so important, they are introduced relatively late in the book for three reasons. Firstly, their mathematical properties are not as straightforward as are those of polynomials, and it seems better to use the mathematically simple polynomial functions to illustrate the principles and methods of the calculus. Secondly, students often have difficulty in assimilating the concept of the number e, but the difficulty is minimized if one can discuss e^x as a function whose gradient is always equal to the function itself. Thirdly, it is convenient to define e by means of the exponential series. Accordingly, the preceding chapter, Chapter 8, deals with mathematical series in their own right, as well as serving as a prelude to the exponential series in Chapter 9. Part of Chapter 8 also serves to round-off the elementary presentation of the calculus with some introductory topics on differential equations. The material of Chapter 10, while specific to plant science, is included in the book as it illustrates so well a biological application of the calculus.

Anyone using this as a textbook may thus follow it through in its natural order. There is one possible exception to this. Chapter 11 can be read at any stage, since very little of it depends on material presented earlier in the book, and some of the ideas now appear in school mathematics syllabuses. The material will be needed at different times in biology courses at different institutions, depending on the applications envisaged.

Mention of applications highlights a particular and ever-present difficulty in teaching mathematics to biologists. Since, for the majority of such students, mathematics is not in itself an interesting subject they are very concerned to see the biological relevance of every mathematical topic discussed. Good biological examples involving *simple* mathematics are very scarce. Biological phenomena are so complex that problems which are not so oversimplified as to be far-fetched require either complicated mathematics, or the application of techniques of probability and statistics. Biological examples are presented whenever possible, but there are sizeable portions of the book which are purely on mathematics without any reference to biology. This is inevitable when developing a theme. For example, the convergence of a series and the idea of a limit have hardly any biological connotations, but these ideas lead on to the differential calculus which does have considerable relevance in the life sciences.

Some exercises are provided at the end of each chapter. On the whole, these are to give practice in the methods presented in the chapter, but, where appropriate, exercises involving direct biological situations are presented.

Finally, it should be stated that the level of mathematical knowledge assumed of the reader is that of GCE 'O' level, or equivalent, but where possible, topics are developed from first principles.

I should like to thank Professor Arthur J. Willis for his careful reading of the script and, as always, the staff of Edward Arnold for their friendly co-operation and assistance.

Llanrhystyd, Aberystwyth
1983   D.R.C.

Contents

1
Why mathematics in biology?

Probably the most significant event that occurred during the rise of man to pre-eminence, from being merely 'another animal', was the development and use of language. At first, language was only spoken, but it did enable relatively large quantities of information to be communicated from one individual to another. More than that, language also provided a means for controlling and monitoring thought itself; thus, language enables concepts to be manipulated independently of the objects to which they refer, and later it becomes possible to think logically without reference to any particular objects at all. Hence, during the history of man, abstract thought became feasible, and from this beginning arose philosophy.

If we had to select an area of application of language that has been outstandingly successful, we should undoubtedly choose the expression of human emotion. The evidence for this is clear when one considers the achievements of oratory and literature. However, when it comes to conveying scientific information, ordinary language is less successful. This is because it is almost impossible to convey precise meaning, since most words in a language have more than one meaning, even if these meanings differ only marginally. Again, because of these various shades of meaning, a particular word means a slightly different thing to different people. This is fine for the literary use of a language where part of the onus for interpretation lies with the reader or listener, but it is not so good for scientific use, where data and hypotheses must be presented with complete unambiguity.

A long time, perhaps many thousands of years, after the rise of language, a new kind of medium for transferring information from one individual to another was evolving. This was at the time when the first of the ancient civilizations – Egypt and Sumer – were flourishing. These civilizations, besides being the architects and builders of large buildings and other monuments, also developed a calendar. Such achievements required precision: precision in measurement, precision in the subsequent manipulation of measurement, and precision in the transmission of such data to other people. All this could be achieved through another kind of language – the language of mathematics.

The mathematical language is just the opposite of ordinary language in that its elements are precise and unambiguous, or at least should be. Every quantity and symbol used can be accurately defined in terms of earlier quantities and symbols already defined. Thus, mathematics is built up on precision and fact, whereas ordinary languages are, to some extent, based on the variability and imprecision of human feelings and emotions.

The development of a science

The study of any science, viewed historically, consists of two main phases. First it is studied almost exclusively from a qualitative point of view; but after an initial period, quantitative methods come to be used increasingly. One of the main reasons for this type of development, from the qualitative to the quantitative, is that a science begins as an observational study, progresses to an experimental, and finally to a theoretical study. At first, phenomena are observed as they occur in nature; later scientific work consists of performing experiments, drawing inferences from the results, and then trying to formulate general laws. By their nature, some sciences have to make the jump from observation to theory without the intervening benefits of experimentation. Astronomy is an obvious example here, and it is remarkable that observational and theoretical astronomy have proceeded side by side for several millennia.

At the present time, physics is the science which is pre-eminent in the use of mathematics. Many physical phenomena are rather less complex than are those of other sciences, and the subject has progressed through the stages of observation and experiment, and has emerged as a theoretical science. This is not to say that observation and experiment are not still carried out in physics; the main point is that physics has reached the stage in which there exists a substantial body of theory, mathematical in nature, which has its origins in observation and experiment. In physics at the present time, the experimental and theoretical sides of the subject are of equal importance.

The phenomena of chemistry are often more complex than those of physics, and this subject has not progressed quite as far as physics on the theoretical side. Although there have been spectacular advances in theoretical chemistry in recent years, chemistry as a whole is still, at present, somewhat more of an experimental science than is physics.

In biology, the situation is very different. Firstly, biological phenomena are highly complex; secondly, there is almost unlimited scope for pure observation of biological phenomena. Hence it is mainly only in the present century that biology has become an experimental science, whereas experimental work in the physical sciences has been undertaken for several hundred years. As a result, it is only now that 'theoretical biology' is tentatively emerging.

From the above remarks, it would seem that mathematical theory is the ultimate aim in a scientific discipline. This is true; not for its own sake, but because in the last resort the phenomena of nature can only be explained in the precise terms of mathematics. Consider this example from physics.

Observation: a stick immersed in water at an angle to the surface (other than a right-angle) so that part of the stick is out of the water and part submerged; the stick appears bent at the surface of the water.

Experiment: a vessel of water is set up on the laboratory bench, and rays of light are traced through the water for various angles of the incident beam; it is found that at an air–water surface (assume that the vessel is made of very thin glass) the ratio of the sine of the angle of the light beam on the air side of the surface to the sine of the angle of the beam on the water side of the surface is constant, and this constant ratio is called the refractive index.

Theoretical deductions: this experimental result can be used in conjunction with facts gleaned from experiments on other phenomena of light, such as reflection, diffraction, interference, to establish knowledge on the nature of light. For instance, it has been found that the velocity of light in a dense medium is less than in a sparse one. This latter experimental finding coupled with the result of the refraction experiment can be analysed mathematically to show that light travels in a wave form.

For this particular example, there is an obvious relationship between observation, experiment, and theoretical deduction. Such examples can be multiplied many fold. In chemistry, we observe a particular reaction, and we experiment to find out the exact conditions under which the reaction occurs. When we then enquire why this particular reaction occurs and not some other, it is necessary to look to the concepts of physical and theoretical chemistry, both of which are founded on mathematics.

Whether or not all biological phenomena can be explained by the physical sciences, or that ultimately it is found that the property of life is 'something extra', it is already quite evident that the manifestations of 'life' can be explained in terms of the physical sciences, particularly chemistry. Since the physical sciences are based on mathematics, so also, indirectly, are the biological sciences.

In summary, experimental results are usually in a quantitative form, even in biology, and therefore sound theoretical deductions can normally only be made by mathematical analysis. This is why, ultimately, mathematics is indispensable to any science; and so any scientist, whatever his or her speciality, should have an adequate knowledge of mathematics.

Biology, mathematics, and statistics

The mathematical model

From the penultimate paragraph of the previous section, one might infer that the utility of mathematics to the biologist is indirect, arising only after experimental results have been interpreted by the concepts of physical science. This, however, is not so. Mathematics is applied directly to the results of biological observation and experiment in a similar manner to the physical sciences but, because of the complexity of the phenomena, its application is much more difficult.

In the present state of biological knowledge, it is impossible to apply a rigorous mathematical analysis to a biological system, such as may be applied, for example, to an electric circuit. What is done, however, is to construct a ***mathematical model*** of the phenomenon in which we are interested. Certain

assumptions about the system have first to be made, and put into mathematical form. These assumptions are based on current knowledge obtained from previous observations and experiments. Next, appropriate **mathematical methods** are applied to the assumptions to achieve an end result which **simulates** the system under study. The simulated result can then be compared with what actually happens. If agreement between the theoretical result and the observed happening is good, then we gain further insight into the process under study; and moreover, we can use the model for predictive purposes. In any science, an ultimate aim is **prediction.** For instance, in an electrical circuit we can predict how the current will change for a given change of voltage, using a simple mathematical model of the circuit (Ohm's Law). In a biological system that has been 'described' mathematically by means of a model, predictions of what will happen under certain changes of conditions can also be made. If the results of using the model do not agree with actuality, then one or more of our basic assumptions must be wrong (assuming the absence of mathematical errors!), and so, in a negative sense, our knowledge is still increased. An example of the construction of a very simple mathematical model is given in Chapter 6.

Statistics in biology

There is yet another complication to the would-be user of quantitative methods in biology, and that is **variability.** The phenomenon of variability is not confined to biology, but arises whenever experimental work is undertaken. Even in the physical sciences, repetitions of a single experiment will give slightly different results, e.g. measurement of the refractive index of a substance, or the location of an end point in volumetric analysis. This kind of variability, which is called **experimental variability** or **experimental error**, arises solely because a human being attempts to measure something; the something does not change, but the reactions of the human being during the conduct of the experiment do change.

Experimental variability also occurs in biology, but here it is considerably augmented by the variability inherent in biological material. If we measure the refractive index of a block of glass very carefully, we are safe in asserting that our result *is* the refractive index of this kind of glass, under the conditions of the experiment. On the other hand, if we measure the increase in height of a sunflower plant over one day, we certainly cannot say that this is the growth rate of sunflower plants in general, even under the same experimental conditions. The same plant may have a noticeably different growth rate at an earlier or later stage in its growth; and even if we take two plants which germinated from the same source of seed at the same time, they will almost certainly show different growth rates at any instant, aside from experimental error. So to be able to make any sort of general statement about the growth rate of sunflower plants of a given age and under defined conditions, we have to measure several plants and take an average. This immediately raises the question as to how reliable our result is, and this cannot be answered without employing a branch of mathematics known as **statistics**.

Even when no experiment is involved and one is only trying to summarize observations usefully and build a model from them, a simple mathematical approach may not be very satisfactory because of the variability of biological material. A good illustration is afforded by *example 9.1* on page 154. Read the general description of the situation through, note that a mathematical expression is used to describe the situation, and then carefully read the questions asked, each one of which obviously requires a single numerical answer. Now, without worrying about how the answers were obtained, read the last sentence of each of the three sections, and note that each answer is a precise figure. Bearing in mind that a 'cohort' in this context is a natural stand of similar-aged plants, it is quite obvious that these precise answers are only statements of likely results around which actual results will deviate to a greater or lesser extent. One immediately asks, 'How much deviation can be expected?'. The **deterministic** mathematical model that has been erected to describe the situation in this example cannot answer such a question. If, however, the same model had been set up, but with an added feature – a **probability** structure – then we should be in a position to answer questions like the above. The mathematical model would now be a **stochastic** model; it is much more realistic, and more complex.

Both the mathematics of stochastic models, and of statistical methods for the analysis of experiments, are based upon the same theoretical subject – probability and statistics. It is a branch of applied mathematics in the broad sense, not in the narrow sense that the term 'applied mathematics' is often used to denote applications to physics. Therefore, probability theory and statistical science are based on mathematics, and a good knowledge of the subject is necessary for their study. In this book, we shall not deal with probability and statistics. Our concern will be with such topics in mathematics that are of a general nature, topics that have direct biological relevance, and also those that form a basis for the study of statistical science about which the modern biologist needs to know.

2
Numbers, indices and logarithms

Broadly, this chapter is concerned with numbers and number systems. Numbers, including those defined by symbols (letters), are fundamental to mathematics, and so this chapter should be carefully read and understood even if you find that much of it is revision of material already familiar.

Numbers

Imagine a straight line, as in Fig. 2.1, extending indefinitely in each direction. The centre of this line will represent zero, and then at equal intervals on either side of the zero we may mark off points which represent the whole numbers: 1, 2, 3; -1, -2, -3, etc. By convention, positive numbers increase from zero to the right, and negative numbers increase from zero to the left. It is important to note that the symbols $+\infty$ and $-\infty$ do not represent numbers, however large. These symbols may be interpreted in different ways according to their context. Here, they mean that the line extends each way indefinitely.

Fig. 2.1   The real number scale.

Real numbers

Any number on the line defined above and shown in Fig. 2.1 is known as a ***real number***; in other words, such numbers can be represented physically on a scale. Real numbers are sub-divided further, as follows.

   Integer   An integer is a whole number, such as 3, 8, -45, 501.

   Rational number   A rational number is one that can be expressed as the quotient (or ratio, hence the name) of two integers. Thus all whole numbers are rational, since each is the quotient of itself and 1. Also, many non-integers are rational, such as 1.5 $(=\frac{3}{2})$, $2.\dot{3}$ $(=\frac{7}{3})$ and -1.8 $(= -\frac{9}{5})$. A dot over a figure to the right-hand side of the decimal point indicates that that figure recurs indefinitely.

   Irrational number   An irrational number is one that cannot be expressed as the quotient of two integers. Examples are $\sqrt{2}$, $\sqrt{7}$, and π. In general, a number which is neither a terminating nor a recurring decimal is an irrational number.

Thus, a real number may be either rational or irrational; and, if rational, may be either an integer or a fraction.

It may be wondered just why it is necessary to classify the real numbers in this way. The three types of real numbers that we have just considered have evolved historically in the same way. The idea of number originated in the counting of objects, giving rise to integers. The simple arithmetic processes of addition, subtraction, and multiplication of integers always yield other integers. However, the division of one integer by another does not necessarily give a further integer; so, in order to give meaning to the arithmetic operation of division, another type of number besides the integer has to be visualized. This new kind of number is the rational number; and since whole numbers *can* result by dividing one integer by another, then rational numbers must include integers. When a rational number is a fraction, it either terminates (as in $\frac{3}{2} = 1.5$) or recurs (as in $\frac{7}{3} = 2.\dot{3}$). Recurrence is not necessarily confined to one figure: it may be a whole group of figures. Thus

$$\tfrac{6}{7} = 0.85714285714285714$$

correct to 17 decimal places, so we can write $\frac{6}{7} = 0.\dot{8}5714\dot{2}$ since this group of figures recurs indefinitely. A rational number can be written down accurately either as a vulgar fraction or as a decimal; even $0.\dot{8}5714\dot{2}$ is an accurate *representation* of $\frac{6}{7}$.

If we now return to the integers, squaring means multiplying an integer (or any number) by itself, e.g. $3 \times 3 = 3^2 = 9$. But the opposite process, taking a square root, may give a number which is not an integer, a terminating fraction, or a recurring fraction, i.e. it is not a rational number. To give meaning to the square root, another class of numbers must be designated – the irrational numbers. The main feature of these numbers is that they cannot be accurately written in fraction or decimal form. For instance, $\sqrt{2} = 1.414213562$ to 9 decimal places; and $\pi = 3.14159$ correct to 5 decimal places, or we can write $\pi \simeq \frac{22}{7}$. The sign $\simeq$ means 'approximately equal to' and it is used here to show that $\frac{22}{7}$ *might* be used in place of π in numerical work. How good the approximation is may be ascertained by evaluating $\frac{22}{7}$ in decimal form, which is $3.\dot{1}4285\dot{7}$; and it is evident that $\frac{22}{7}$ differs from π by approximately 0.00127 – a difference of about 0.04% of the quantities under discussion. The decision as to whether a particular approximation of an irrational number is adequate can only be judged by prevailing circumstances, but the only ways in which irrational quantities can accurately be referred to are in the forms $\sqrt{2}$ and π, for example.

Complex numbers

Consider the following quadratic equation:

$$x^2 + 2x + 2 = 0 \tag{2.1}$$

i.e. $$(x + 1)^2 + 1 = 0$$

or $x + 1 = \pm\sqrt{-1}$

Hence $x = -1 \pm \sqrt{-1}$

Hitherto you have probably been told that $\sqrt{-1}$ does not exist. This is perfectly true if the statement is taken to mean that $\sqrt{-1}$ is not a real number. However, a cursory glance at *equation 2.1* does not reveal anything peculiar about it; it is a perfectly normal quadratic equation, but it gives a solution containing a quantity which is not a real number. Now although we can imagine the existence of (say) two objects such as loaves of bread, $\frac{1}{2}$ a loaf, or even $\sqrt{2}$ of a loaf, the mind boggles at the thought of $\sqrt{-1}$ of a loaf! This quantity cannot be perceived simply because it is not a real number, and it does not exist on our scale of real numbers as shown in Fig. 2.1.

The mathematician does not, however, dismiss *equation 2.1* as an impossible type; he invents a new class of numbers and calls them **complex numbers**. The square root of minus one is known as an **imaginary number**, and is always denoted by *i*. A complex number has the form $a + ib$, where a and b are real numbers, and this is precisely the form that the roots (see page 46) of *equation 2.1* take (here $a = -1$ and $b = 1$ or -1).

Although complex numbers are a creation of the mathematician's mind, they are extremely useful in the solution of practical problems. However, the subject is beyond the scope of this book, and we shall restrict ourselves entirely to real numbers.

The factorial of a positive integer

The factorial of an integer, n, is usually designated as $n!$ (or occasionally as $\lfloor n$), and is defined as the product of n and all preceding integers down to 1;

i.e. $n! = n(n - 1)(n - 2)\ldots(2)(1)$ (2.2)

For example $4! = 4 \times 3 \times 2 \times 1 = 24$

and $6! = 6 \times 5 \times 4 \times 3 \times 2 \times 1 = 720$

Now although $n!$ represents a real number, an integer in fact, it does not rank with the types of number that we have been discussing above, e.g. rational and irrational numbers. Rational numbers, for example, are a natural sub-class of real numbers, and they have only been given a special name to make it easy to think of them as a sub-class on their own. The factorial of an integer, on the other hand, is an example of mathematical notation. The product represented by the factorial of an integer occurs often in mathematics, and so the mathematician defines this product as we have already done, and gives it a name (factorial) and a symbol (!). Mathematics abounds with specialized notations and many will be introduced throughout this book.

Let us consider two properties of factorials. Firstly

$$n! = n(n - 1)!$$ (2.3)

e.g. $6! = 6 \times (5 \times 4 \times 3 \times 2 \times 1) = 6 \times 5!$

This property is useful in cases where we need to evaluate factorials of successive integers; there is no need to multiply right down to 1 each time. For instance, we already know that $6! = 720$;

hence $7! = 7 \times 6! = 7 \times 720 = 5040$

$8! = 8 \times 7! = 8 \times 5040 = 40\ 320$ etc.

Evidently $n!$ increases very rapidly as n increases.

The second property of factorials show that the factorial notation can be used to denote the product of a set of successive integers even if they do not extend down to 1. Consider

$$7 \times 6 \times 5 = \frac{7 \times 6 \times 5 \times 4 \times 3 \times 2 \times 1}{4 \times 3 \times 2 \times 1} = \frac{7!}{4!}$$

The product $4 \times 3 \times 2 \times 1$ appears in both the numerator and the denominator of the fraction, and so cancels out. The above example can be generalized. Suppose we have two integers, n and r, and that r lies *between* 1 and n, i.e. $1 < r < n$ (1 is less than r, and r is less than n). Then

$$n(n-1)\,(n-2)\ldots(r) = \frac{n(n-1)\,(n-2)\ldots(1)}{(r-1)\,(r-2)\ldots(1)} = \frac{n!}{(r-1)!} \qquad (2.4)$$

Finally, note that factorials of numbers other than positive integers can be defined, but this requires mathematical theory beyond that presented in this book (*see A Biologist's Advanced Mathematics*). There is, however, one result of this theory that is important in elementary biomathematics, and that is

$$0! = 1 \qquad (2.5)$$

I am afraid that this curious-looking statement will just have to be accepted.

Indices

A number of the form a^m is defined as the number a raised to the power m; a is usually called the base, and m the index, power, or exponent. For the moment, we assume that m is a positive integer; then a^m means that a is multiplied by itself m times. If we have an expression of the form $a^m \times b^n$ wherein the bases of the two numbers are different, then the expression cannot be simplified further; but if the bases are the same number, we may establish three laws. We shall do this by means of specific examples, and so the table of values below will be useful.

$2^1 = 2$	$3^1 = 3$
$2^2 = 4$	$3^2 = 9$
$2^3 = 8$	$3^3 = 27$
$2^4 = 16$	$3^4 = 81$
$2^5 = 32$	$3^5 = 243$
$2^6 = 64$	$3^6 = 729$

Law 1. Multiplication

$$\text{Example} \qquad 2^2 \times 2^3 = 4 \times 8 = 32 = 2^5 = 2^{(2+3)}$$

$$\text{Example} \qquad 3^1 \times 3^5 = 3 \times 243 = 729 = 3^6 = 3^{(1+5)}$$

$$\text{So, generally} \qquad a^m\, a^n = a^{(m+n)} \qquad\qquad (2.6)$$

Law 2.　Division

$$\text{Example} \qquad 2^5/2^3 = 32/8 = 4 = 2^2 = 2^{(5-3)}$$

$$\text{Example} \qquad 3^3/3^2 = 27/9 = 3 = 3^1 = 3^{(3-2)}$$

$$\text{So, generally} \qquad a^m/a^n = a^{(m-n)} \qquad\qquad (2.7)$$

Law 3. Powers of indices

$$\text{Example} \qquad 2^3 \times 2^3 = (2^3)^2 = 8^2 = 64 = 2^6 = 2^{(3\times2)}$$

$$\text{Example} \qquad 3^2 \times 3^2 = (3^2)^2 = 9^2 = 81 = 3^4 = 3^{(2\times2)}$$

$$\text{So, generally} \qquad (a^m)^n = a^{mn} \qquad\qquad (2.8)$$

The above three laws should be familiar to you already, and have been derived assuming that m and n are positive integers. We now assume that these laws are valid for all values of m and n; thus indices may be positive or negative integers, zeros, or fractional numbers. Fractional numbers may by positive or negative, and rational or irrational. We therefore need to find meanings for the expressions a^0, a^{-m}, and $a^{m/n}$.

Theorem 2.1　　The value of a^0 is 1

$$a^0 = a^{(m-m)} = a^m/a^m \quad \text{(Law 2). But} \quad a^m/a^m = 1.$$

$$\text{Thus} \qquad a^0 = 1 \qquad\qquad (2.9)$$

Theorem 2.2　　The value of a^{-m} is the reciprocal of a^m

$$a^m\, a^{-m} = a^{\{m+(-m)\}} = a^{(m-m)} = a^0 \quad \text{(Law 1)}$$

But $a^0 = 1$ (Theorem 2.1); hence $a^m\, a^{-m} = 1$.

$$\text{Therefore} \qquad a^{-m} = 1/a^m \qquad\qquad (2.10)$$

Theorem 2.3   The value of $a^{m/n}$ in the nth root of a^m

$$(a^{m/n})^n = a^m \quad \text{(Law 3)}$$

Taking the nth root of both sides, we have $a^{m/n} = \sqrt[n]{a^m}$ $\hspace{2cm}$ *(2.11)*

The argument of Theorem 2.3 may be clarified by a simple example:

Example 2.1
  Evaluate $a^{1/2}$.

$$(a^{1/2})^2 = a^{\{(1/2) \times 2\}} = a^1 = a$$

$$\text{i.e.} \quad (a^{1/2})^2 = a$$

Taking the square root of both sides, we have

$$\sqrt{\{(a^{1/2})^2\}} = \sqrt{a}$$

But the square root of a squared number (i.e. the left-hand side) is the number itself, so

$$a^{1/2} = \sqrt{a}$$

Example 2.2
  Evaluate $5^{-2} + 3^{2/3} - 2^{-1/2} + 4^{-3/2}$

The expression simplifies to $\hspace{1cm} \dfrac{1}{5^2} + \sqrt[3]{3^2} - \dfrac{1}{\sqrt{2}} + \dfrac{1}{\sqrt{4^3}}$ $\hspace{1cm}$ (Theorem 2.3)

$$= \frac{1}{25} + \sqrt[3]{9} - \frac{1}{\sqrt{2}} + \frac{1}{8}$$

$$\simeq 0.04 + 2.0801 - 0.7071 + 0.125\text{*}$$

* This decimal expression is only approximately equal to the above common fraction expression, since $\sqrt{2}$ and $\sqrt[3]{9}$ are irrational.

$$\text{Therefore} \hspace{1cm} 5^{-2} + 3^{2/3} - 2^{-1/2} + 4^{-3/2} = 1.538$$

(correct to 3 decimal places)

Logarithms

Previously, your prime concern with logarithms has been as a means to simplify the arithmetic of multiplication and division. Logarithms are, however, very important in mathematics for other reasons, and they are of particular relevance in quantitative biology. Hence we now need to study some aspects of the theory of logarithms.

Definition ***The logarithm of a number to a certain base is the power to which the base must be raised to give the number***

$$\text{Thus if} \qquad \log_a m = x \qquad\qquad (2.12)$$

$$\text{this implies that} \qquad a^x = m \qquad\qquad (2.13)$$

The base is vital. A logarithm cannot be defined unless the base is specified: conversely, the logarithm of a number to one base is different from the logarithm of the same number to another base (e.g. $\log_{10} 2 = 0.3010$ but $\log_2 2 = 1.0000$). It should be noted that the base of a logarithm may be any positive real number; note also that we define the logarithm of a positive number only.

Theorem 2.4   The logarithm of the product of two numbers is equal to the sum of the logarithms of the numbers

$$\text{i.e.} \qquad \log_a (mn) = \log_a m + \log_a n \qquad\qquad (2.14)$$

$$\text{Let} \qquad \log_a m = x \qquad \text{and} \qquad \log_a n = y$$

$$\text{then} \qquad a^x = m \qquad \text{and} \qquad a^y = n$$

$$\text{Now} \qquad mn = a^x a^y = a^{(x+y)} \quad \text{(Law 1)}$$

Hence, on invoking *relationships 2.12* and *2.13*, we have

$$\log_a (mn) = x + y = \log_a m + \log_a n$$

Theorem 2.5   The logarithm of the quotient of two numbers is equal to the difference of the logarithms of the numbers

$$\text{i.e.} \qquad \log_a (m/n) = \log_a m - \log_a n \qquad\qquad (2.15)$$

The proof follows the reasoning in Theorem 2.4, but Law 2 of indices is used instead of Law 1.

Theorem 2.6   The logarithm of a number raised to a power is equal to the power multiplied by the logarithm of the number

$$\text{i.e.} \qquad \log_a (m^n) = n \log_a m \qquad\qquad (2.16)$$

$$\text{Let} \qquad \log_a m = x \qquad \text{then} \qquad m = a^x$$

$$\text{So} \qquad m^n = (a^x)^n = a^{xn} \quad \text{(Law 3)}$$

Hence, on invoking relationships *2.12* and *2.13*, we have

$$\log_a (m^n) = xn = n \log_a m$$

The above three theorems define the theory that enables logarithms to simplify arithmetical computation, by converting multiplication into an addition procedure, division into subtraction, and raising a number to a power into a multiplication process. As regards the last statement, which is associated with Theorem 2.6, it is evident that $n \log_a m$ is much more tractable than m^n. Indeed, unless n is a rational number, m^n can only be defined and evaluated (approximately) *via* the form $n \log_a m$ (see page 15).

Common logarithms

Before the universal availability of electronic calculators, logarithms to the base 10 were used in arithmetic calculations, and were widely tabulated. Logarithms to base 10 are thus called **common logarithms**, and the short table below shows why these logarithms are the most convenient in calculations.

$$\begin{array}{llll}
\text{Since } 10^0 = & 1 & \text{then } \log_{10} 1 & = 0 \\
10^1 = & 10 & \log_{10} 10 & = 1 \\
10^2 = & 100 & \log_{10} 100 & = 2 \\
10^3 = & 1000 & \log_{10} 1000 & = 3 \text{ etc.}
\end{array}$$

Hence all that is required is a table of logarithms to the base 10 of numbers between 1.0 and 9.9999, in which the corresponding logarithmic values lie between 0 and almost 1. The logarithm of any number $\geqslant 10$ will be given by using the table and inserting the appropriate integer in front of the decimal point. The integer part of a logarithm is known as the characteristic, and the decimal part as the mantissa.

From a mathematical viewpoint, common logarithms are of little importance; and now that their employment in arithmetic operations has been rendered obsolete by the pocket calculator, they should hardly ever need to be used. However, scientific calculators always provide this function using the key marked 'log'.

Antilogarithms

$$\text{If} \qquad \log_{10} y = x$$

$$\text{then} \qquad 10^x = y$$

(*equations 2.12* and *2.13*). Thus if we know the value of x and wish to know what y is, we require the antilogarithm of x. For common logarithms this is simply 10^x. Anti-common logarithms are also provided for on a scientific calculator using the key marked '10^x'.

Natural logarithms

Logarithms which arise in mathematical work, other than arithmetic, have a base which is an irrational number universally symbolized as e. Its value is approximately 2.71828. It may well be asked why such a peculiar number should give rise to logarithms which are called 'natural' and have simple mathematical properties. The reasons for this will become apparent later in this book (Chapter 9), and the number e will be derived from first principles and its properties described (page 151).

It should be noted that $\log_e$ is often abbreviated to ln (remember this as '*log natural*').

$$\text{Thus} \quad \log_e m = \ln m \tag{2.17}$$

and it is the 'ln' symbol that is used on pocket calculators, with 'e^x' as the anti-natural logarithm.

Changing the base of a logarithm

Normally, logarithms are only tabulated to the bases of 10 and e. Occasionally, however, we require the log of a number to some other base; for example, logarithms to the base of 2 are used in formulae for species diversity in ecological work. Such logarithms can be evaluated from common logs or natural logs by using a theorem called the 'change of base theorem'.

Theorem 2.7 The logarithm of a number to a base, *a*, is equal to the logarithm of the same number to another base, *b*, divided by the logarithm of *a* to the base *b*

$$\text{That is} \quad \log_a m = \frac{\log_b m}{\log_b a} \tag{2.18}$$

$$\text{Let} \quad \log_a m = x$$

$$\text{then} \quad a^x = m \tag{2.19}$$

Take logarithms to base *b* of both sides of 2.19

$$\log_b (a^x) = \log_b m$$

$$\text{i.e.} \quad x \log_b a = \log_b m \quad \text{(Theorem 2.6)}$$

$$\text{and so} \quad x = \frac{\log_b m}{\log_b a}$$

$$\text{But} \quad x = \log_a m, \quad \text{hence} \quad \log_a m = (\log_b m)/(\log_b a)$$

Example 2.3

Evaluate $\log_2 7$.

From a calculator we find that $\log_{10} 7 = 0.8451$ and that $\log_{10} 2 = 0.3010$.

$$\text{Hence} \qquad \log_2 7 = \frac{\log_{10} 7}{\log_{10} 2} = \frac{0.8451}{0.3010} = 2.873$$

Example 2.3 shows that the logarithm to base 2 of any number can be obtained by dividing the common log of that number by 0.3010, or alternatively by multiplying by the reciprocal of 0.3010 ($=3.322$).

$$\text{Therefore} \qquad \log_2 m = 3.322 \times \log_{10} m$$

Similar 'conversion factors' may be evaluated to obtain logs to any base from logs to base 10.

The logarithm as the definition of a number raised to any power

A number such as 3^2 is perfectly comprehensible: it is 3 multiplied by itself, $3^2 = 3 \times 3 = 9$. Similarly, 3^{-2} is understandable: $3^{-2} = 1/3^2 = 1/9 = 0.\dot{1}$. Further, $3^{3/2}$ is interpreted as $\sqrt{(3)^3} = \sqrt{27} \simeq 5.196$. But what about such a number as $3^{2.1}$, or even 3^π?

Example 2.4

Evaluate (*a*) $3^{2.1}$ and (*b*) 3^π.

(*a*) This number can be expanded:

$$3^{2.1} = 3^{(2 + 1/10)} = 3^2 \times 3^{1/10} = 3^2 \sqrt[10]{3} = 9 \sqrt[10]{3}$$

No calculator can directly evaluate the tenth-root, so this part of the expression would have to be evaluated by logarithms. Splitting 2.1 up has not therefore made our task any easier: it is much the simplest plan to keep the power in its original form, and use logarithms throughout.

$$\text{Put} \qquad x = 3^{2.1}$$

Take logarithms (to any base) of both sides:

$$\log_a x = 2.1 \, (\log_a 3) \quad \text{(Theorem 2.6)}$$

Common logs are convenient, so

$$\log_{10} x = 2.1 \, (\log_{10} 3)$$
$$= 2.1 \times 0.4771$$

$$\text{i.e.} \qquad \log_{10} x = 1.0019$$

To find x itself $(=3^{2.1})$ we need the antilog of 1.0019, which is 10.04.

$$\text{Therefore} \qquad 3^{2.1} \simeq 10.04$$

(*b*) Put $x = 3^{\pi}$

Take common logs of both sides: $\log_{10} x = \pi \log_{10} 3$
$$= 0.4771\pi$$

To evaluate this further we must give π an approximate value, which can be 3.142 correct to 3 decimal places.

$$\text{Hence} \qquad \log_{10} x \simeq 1.499 \qquad \text{and so} \qquad x \simeq 31.55$$

$$\text{Therefore} \qquad 3^{\pi} \simeq 31.55$$

Any number of the form a^n can be evaluated by the procedure illustrated in *example 2.4* provided $a > 0$, and if n is irrational this is the *only* way in which a^n can be defined and evaluated. If n is rational, a^n can be thought of more simply in terms of powers and roots, but one may still have to use logarithms to evaluate it. It should also be noted that taking the logarithm of a number usually gives only an approximate result. Hence, both answers of *example 2.4* are approximations.

Many scientific calculators have an 'x^y' key to perform the operation of raising a number to a power, but the method used is precisely that shown in *example 2.4* involving logarithms.

EXERCISES

1. Evaluate the factorials of all the integers from 1 to 10.
2. Find the values of (*a*) 10!/8! and (*b*) 12!/(9!3!).
3. Evaluate $0.001529^{2/9}$.
4. Show that $(\log_a b)(\log_b a) = 1$.
5. Prove that (*a*) $\log_a a = 1$ and (*b*) $\log_a a^x = x$.

3

Multi-dimensional space

At first sight, the idea of multi-dimensional space sounds strange and forbidding, more connected with modern physics and cosmology than with biology. In fact, however, the concept of multi-dimensional space is not difficult to comprehend, and considerable biological use is made of the ideas involved, especially in the fields of ecology, genecology, and taxonomy.

Location in space

Cartesian co-ordinates in two dimensions

Let us begin by considering the familiar two-dimensional space which can be represented on the plane of the paper, and consider first how to specify the position of a point on the plane. Any specification of the position of a point can only be given *relative* to the position of something else. It is most convenient to make this 'something else' a reference body of some sort, and then measure the distance of the point we require relative to this reference body. Since we are at present working in two dimensions, we require one part of the reference body to measure distances in one direction (say east–west), and another part of the reference body to measure distances at right-angles to the previous direction (i.e. north–south).

The above specification is achieved by drawing two lines at right angles to one another, as shown in Fig. 3.1. These lines are known as the ***co-ordinate axes***, and the horizontal line is usually called the ***x-axis*** while the vertical line is known as the ***y-axis***. The position of any point on the plane may then be specified by two quantities – a distance measured along the x-axis, often called the ***abscissa***, and a distance measured along the y-axis, often known as the ***ordinate***. In Fig. 3.1 the point P is specified by two numbers: X, representing a definite distance along the x-axis (abscissa), and Y, which represents a definite length along the y-axis (ordinate). These two numbers, together with the name or symbol of the point in question, are specified thus: $P(X, Y)$ and are called the ***Cartesian co-ordinates*** of point P, after Descartes who first specified the position of a point in this way.

The co-ordinates of the point of intersection of the axes are (0,0); this point is known as the ***origin***, and is designated by 0. You will doubtless recognize the co-ordinate axes as the same as those used in elementary graphical work, but

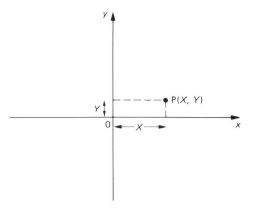

Fig. 3.1   The Cartesian co-ordinates of a point in two-dimensional space.

are probably more used to seeing the origin in the bottom left-hand corner of the graph instead of in the middle, as in Fig. 3.1. However, the convention of measuring distances from the origin is the same in both cases: increasing distances to the right of the origin are associated with increasing positive values of x, and increasing distances above the origin are associated with increasing positive values of y. What about points to the left of and/or below the origin? Working away from the origin along the x-axis on the left-hand side we encounter increasing *negative* values of x, and as we move down from the origin along the y-axis we find increasing negative values of y.

The statements in the previous paragraph are illustrated in Fig. 3.2. In these diagrams, the intersection of the x- and y-axes at right angles delimit four quadrants, which can be named: 'upper-right', 'upper left', 'lower-left', and 'lower-right'. Figure 3.2(a) shows four points, each one being situated 2 units along the x-axis and 1 unit along the y-axis, in various directions. It can be seen that in the upper-right quadrant defined by the co-ordinate axes, values of both

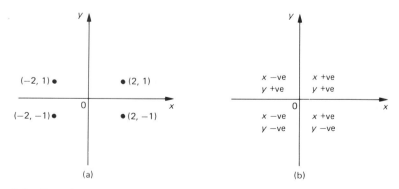

Fig. 3.2   Cartesian co-ordinates in the four quadrants defined by co-ordinate axes in two-dimensional space: (**a**) particular and (**b**) general.

x and y are positive. In the upper-left quadrant, values of x are negative while those of y are positive. In the lower-left quadrant, both x and y are negative; while in the lower-right quadrant, x is positive and y is negative. These facts are summarized in Fig. 3.2(b).

As biologists, we are mainly interested in quantities which can be depicted in the upper-right quadrant – positive values of x and y. Nevertheless, we must remember that the other three quadrants exist, because, when analytical methods are applied to real data, negative values do sometimes arise.

Three dimensions

All the above ideas extend quite easily to three dimensions, and although the concept of three dimensions is everyday experience, it is difficult to illustrate three-dimensional space on paper. In attempting to do this, we use the laws of perspective and the adaptiveness of the human eye to perceive what it is asked to.

The three dimensions are conveniently termed breadth, depth, and height. If we imagine breadth and depth to be a horizontal plane (x- and y-axes), then the third dimension, height, can be represented by a vertical line (z-axis). This is illustrated in Fig. 3.3, and for simplicity we shall consider only that part of the co-ordinate system in which all quantities are positive. The situation illustrated in Fig. 3.3 can be realized in practice as a bar placed vertically at the corner of a table to represent the z-axis, and then the two edges of the table, beginning from the corner on which the bar is standing, will represent the x- and y-axes. The position of any point P on or above the table can then be specified precisely with respect to these three co-ordinate axes, as shown in Fig. 3.3. From P a perpendicular is dropped on to the x–y plane; the length of this perpendicular line is Z. The point at which the perpendicular intersects the x–y plane has co-ordinates (X,Y). Hence the co-ordinates of point P are (X,Y,Z).

Higher dimensions: the mathematical and physical concepts of dimension

Although a pictorial (i.e. physical) representation of the system of co-ordinates in four or more dimensions is not possible, the *mathematical* idea of a number of dimensions greater than three is perfectly feasible and sound. The concept is entirely a mathematical one and has no physical reality; the fact that we are able to realize 1, 2 and 3 mathematical dimensions in physical space is mere coincidence. The mathematical concept of dimension is distinct from the physical concept in which there is an upper limit on the number of possible dimensions. According to present knowledge the physical universe is composed of four dimensions, three of which are space and one is time; although the dimension of time is clearly different from the other three of space. In the mathematical sense, the number of possible dimensions is infinite.

We have been dealing with the location of points in two- and three-dimensional space, and because these dimensions in the mathematical sense

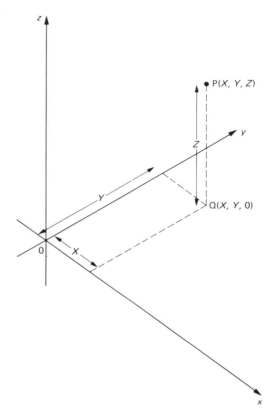

Fig. 3.3   The Cartesian co-ordinates of a point in three-dimensional space.

can be made to accord with physical dimensions, it has been possible to give physical meanings to a set of co-ordinates for a point P, e.g. (X, Y, Z). In two dimensions the location of a point is precisely specified by two co-ordinates; similarly, in three dimensions a point is completely specified by quoting three co-ordinates. Is there any reason why we cannot add more co-ordinates to specify the position of a point in a 'space' of more than three dimensions? No reason at all. There is in fact nothing to stop us from writing the co-ordinates of a point P as say, (X, Y, Z, T, U), and we should then infer that point P was situated in a 'space' of five dimensions.

It will be convenient at this stage to change the labelling of the co-ordinate axes from that used previously. Let us now designate the x-axis as the x_1-axis, the y-axis as the x_2-axis, and the z-axis as the x_3-axis. Then, for example, the co-ordinate axes for five-dimensional 'space' will be designated $x_1, \ldots, x_5$, and the co-ordinates of point P in such a space will then be $(X_1, X_2, X_3, X_4, X_5)$.

In the last two paragraphs the word 'space' has been placed in quotation marks because we are now dealing solely with the mathematical idea of space, which is not equivalent to the familiar physical concept of space. From now on, the quotation marks will be dispensed with, although we shall always be referring to the mathematical concept of space.

In view of the fact that the mathematical idea of space has no parallel with physical space beyond the third dimension, you may be asking just what use can the biologist make of the mathematical concept of space, when he has to deal with real objects which exist in the physical sense. The example in the next section introduces one of the most fruitful applications of the idea of multi-dimensional space to biological enquiry.

The niche of an organism in an ecosystem

An *ecosystem* is a system of interrelationships between organisms, and between organisms and their environment. An ecosystem can be considered at any order of size and complexity, and examples are: a patch of soil in a garden, an area of woodland, or a whole island or country. Descriptively, the *niche* of an organism is the 'position' it occupies in relation to the physical factors operating, and to the other organisms present, in the environment. For instance, a particular species of plant may tolerate only an acid soil and be intolerant of competition by heather, to name but two external factors. Thus the species may find a niche on acid soil which does not carry a stand of heather.

The concept of a niche can be put into more definite quantitative terms. The niche of an organism may be defined as its position in multi-dimensional environmental space, and the mathematical ideas already discussed can be applied to the ecosystem. Each one of the co-ordinate axes represents an environmental factor, either physical or biotic, affecting the species concerned. Let us illustrate with an extremely simplified ecosystem consisting of three factors:

(*i*) average temperature of the air, x_1
(*ii*) average moisture content of the soil, x_2
(*iii*) number of earthworms per unit volume of soil, x_3.

In this environmental space, organisms find their respective niches according to their physiological properties (Fig. 3.4). Organism A, which may perhaps be a species of moss, requires a high temperature, a low soil moisture content, and a reasonable earthworm population. The last feature may indicate a soil having a reasonable base status. Thus, moss A could be expected to occur on dry calcareous soils in the south of England. Moss B, however, requires a lower air temperature, a much damper soil, and a small earthworm population perhaps indicating an acid soil. Moss B, then, may be found on siliceous soils in the west and north of Britain, where the climate is cool and wet.

Two points may be noted about this very over-simplified picture. Firstly, the niche of an organism would not be represented as a *point* in multi-dimensional space. Plants and animals have a tolerance range to environmental and biotic factors; this means that in our three-dimensional environment a niche will be

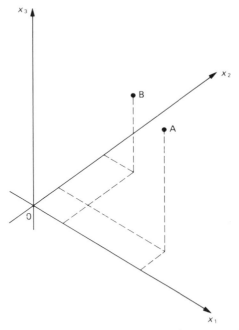

Fig. 3.4 Two points located in three-dimensional space (see page 21).

represented by a *volume* rather than a point. The two-dimensional analogue of a volume is an *area*, while the equivalent in four or more dimensions is known as a *hyper-volume*.

Secondly, in practice the number of environmental factors is indefinitely large. Because of this, and also because we do not know what all the environmental and biotic factors contributing to a niche are, an ecosystem cannot be directly analysed in this way, in the present state of knowledge. The actual procedures used are much too complicated to describe here, but they are based upon the ideas and principles discussed above.

Polar co-ordinates in two dimensions

The Cartesian method of specifying the position of a point relative to co-ordinate axes is only one of several possible methods, although it is the commonest and the easiest to understand. Another system, known as *polar co-ordinates*, is often useful in mathematics, although more rarely in biomathematics. In this system, only one axis is required – the *x*-axis – on which the origin ($x = 0$) is shown, and we require to specify the position of point P in relation to the axis and the origin (Fig. 3.5). The polar co-ordinates of P are given by (r, θ), where r is the length of the straight line from the origin to P, and θ is the angle that this line makes with the *x*-axis. The length r is

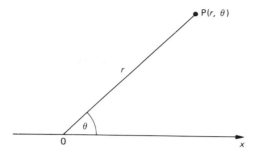

Fig. 3.5 The polar co-ordinates of a point in two-dimensional space.

always positive, and θ can take any value between 0 and 2π radians, i.e. 0° and 360°. The relations between co-ordinates in the two systems are quite simple:

Polar to Cartesian	$X = r \cos \theta$	$Y = r \sin \theta$
Cartesian to Polar	$r = \sqrt{(X^2 + Y^2)}$	$\theta = \tan^{-1}(Y/X)$

The expression 'tan^{-1} (Y/X)' is read as 'the angle whose tangent is Y/X'. An alternative notation is 'arctan'; thus, $\theta = \arctan(Y/X)$.

One use of polar co-ordinates in biological mathematics is concerned with the description of the spiral arrangement of leaves on the stems of plants – phyllotaxis.

Distances between points

Two dimensions

Let $P(X, Y)$ and $Q(X', Y')$ be two points whose distance apart is required (Fig. 3.6a); that is, we want to know the length of the straight line PQ. To find

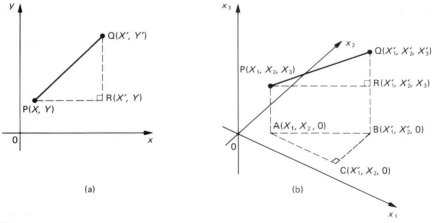

(a)   (b)

Fig. 3.6 The distance between two points: (**a**) in two-dimensional space, (**b**) in three-dimensional space.

out this distance, complete the right-angled triangle PQR as shown in Fig. 3.6(a).

By the theorem of Pythagoras $PQ^2 = PR^2 + QR^2$

The lengths of the lines PR and QR are particularly easy to find. PR is parallel to the x-axis; its length is measured entirely in terms of x, and is $(X' - X)$. Similarly, QR is parallel to the y-axis; its length is measured in terms of y only, and is $(Y' - Y)$.

Hence $PQ^2 = (X' - X)^2 + (Y' - Y)^2$

and the positive square root gives the length of the line PQ.

In order to extend these ideas easily to higher dimensions, let us relabel the axes as x_1 and x_2. The co-ordinates of P will be (X_1, X_2) and of Q will be (X_1', X_2').

Then $PQ^2 = (X_1' - X_1)^2 + (X_2' - X_2)^2$

or, writing $d_1 = (X_1' - X_1)$ and $d_2 = (X_2' - X_2)$

$$D^2 = d_1^2 + d_2^2 \qquad (3.1)$$

The square of the required distance, D^2, is given as the sum of two terms: the first involving a distance on the x_1-axis only, d_1, and the second term involving a distance on the x_2-axis only, d_2.

Three dimensions

Figure 3.6(b) shows the three-dimensional case, which is analogous to the two-dimensional one we have just discussed. Two points, $P(X_1, X_2, X_3)$ and $Q(X_1', X_2', X_3')$ are located in three-dimensional space; what is their distance apart? By use of exactly the same principles as before, it can be shown that

$$PQ^2 = (X_1' - X_1)^2 + (X_2' - X_2)^2 + (X_3' - X_3)^2$$

or $$D^2 = d_1^2 + d_2^2 + d_3^2 \qquad (3.2)$$

although the proof is rather more lengthy than for the two-dimensional case. Try to work it through yourself, starting by dropping perpendicular lines from P and Q to the horizontal plane and completing the right-angled triangle ABC.

The extension from two dimensions (*equation 3.1*), to three dimensions (*equation 3.2*), is immediately apparent. We now have the distance between the two points given by three terms: the first involving a distance measured on the x_1-axis only, the second term involving a distance on the x_2-axis only, and the third term a distance on the x_3-axis only.

n dimensions

The '*D*-formula' is easily extended to *n*-dimensions: it becomes

$$D^2 = d_1{}^2 + d_2{}^2 + d_3{}^2 + \ldots + d_n{}^2 \qquad (3.3)$$

For *n* dimensions the right-hand side of *equation 3.3* consists of precisely *n* terms added together. The first term involves only distances along the x_1-axis (one dimension), the second term involves distances only along the x_2-axis (a second dimension), and so on.

Projection of a distance in *n*-dimensional space into a space of fewer dimensions

It is impossible to draw a graph in four or more dimensions, and coping with three dimensions graphically is not easy. It is possible, however, to gain *some idea* of the positions and distances apart of points in multi-dimensional space by constructing a two-dimensional graph, as we shall now show.

Referring to Fig. 3.6(a) once again, we have already remarked that the length of PR is $(X' - X)$, and that of QR is $(Y' - Y)$. Now imagine viewing the plane edgewise on, at right-angles to the *x*-axis; in other words, the eye is placed at the bottom of the diagram and looking upwards. From this position the eye would not discern the actual distance from P to Q; instead, the length of the line PQ would *appear* to be the length of the line PR. The length PR is called the **projection** of the length PQ on the *x*-axis. Similarly, the length QR is the projection of the length PQ on the *y*-axis. Now unless the line PQ is inclined to both the *x*- and *y*-axes at 45°, these two projected lines will not be of the same length. However, the importance of the projection is that it enables us to represent the distance between the two points which are situated in two-dimensional space (the *x–y* plane) on one dimension only (either the *x*- or the *y*-axis).

Similar reasoning applies to the three-dimensional case in Fig. 3.6(b). Here we can project the distance between P and Q, which are sited in three-dimensional space, into two dimensions: e.g. on to the x_1–x_2 plane wherein the line PR or AB is the projected length, and is equal to $\sqrt{\{(X_1 - X_1)^2 + (X_2' - X_2)^2\}}$. Or, we can go one stage further and project PQ on to a single dimension; either on to the *x*-axis giving a projected length AC of $(X_1' - X_1)$, or on to the x_2-axis giving a projected length BC of $(X_2' - X_2)$, or on to the x_3-axis giving a projected length QR of $(X_3' - X_3)$.

The principal use of projections is to permit us to visualize, in a two-dimensional plane defined by any two of the *n* axes, the orientation and the length of a line joining two points situated in multi-dimensional space. Evidently, the values of the projected distances obtained differ according to which plane the projection is made on. For example, to obtain a full picture of the distance between two points in five-dimensional space, the following 10 graphs would have to be drawn: x_1 against x_2, x_1 against x_3, x_1 against x_4, x_1 against x_5, x_2 against x_3, x_2 against x_4, x_2 against x_5, x_3 against x_4, x_3 against

x_5, and x_4 against x_5. Analytical methods employed, however, ensure that most of the information concerning distances between points is contained in only two or three of the above combinations of axes (see the method of Principal Component Analysis in *A Biologist's Advanced Mathematics*).

Example 3.1

What is the distance between the following pairs of points?

$$(a) \; P(3, 5) \text{ and } Q(4, 7) \quad (b) \; A(2, 3) \text{ and } B(-3, 2).$$

(a) $PQ^2 = (4 - 3)^2 + (7 - 5)^2 = (1)^2 + (2)^2 = 5$

$$\text{Hence} \quad PQ = \sqrt{5} \text{ units}$$

(b) $AB^2 = \{(-3) - (2)\}^2 + \{(2) - (3)\}^2 = (-5)^2 + (-1)^2 = 26$

$$\text{Hence} \quad AB = \sqrt{26} \text{ units}$$

Example 3.2

Two points, P and Q, are sited in five-dimensional space, and their co-ordinates are (3, 5, 4, 3, 6) and (7, 8, 1, 1, 2), respectively. Find the distance between the points. Draw a graph to show the positions of the two points with respect to axes 1 and 2. What is the projected distance between the two points on the plane defined by axes 1 and 2?

First, the distance PQ is calculated as

$$\begin{aligned} D^2 &= (3 - 7)^2 + (5 - 8)^2 + (4 - 1)^2 + (3 - 1)^2 + (6 - 2)^2 \\ &= (-4)^2 + (-3)^2 + (3)^2 + (2)^2 + (4)^2 \\ &= 16 + 9 + 9 + 4 + 16 = 54 \end{aligned}$$

$$\text{Hence} \quad D = \sqrt{54} \text{ units}$$

To plot the graph showing the points projected on to the plane defined by axes 1 and 2, we require the first two co-ordinates of each point, i.e. (3, 5) and (7, 8). The projected distance is then $\sqrt{\{(3 - 7)^2 + (5 - 8)^2\}} = \sqrt{(16 + 9)} = \sqrt{25}$ units (Fig. 3.7).

Quantitative characters in taxonomy

Examples of the use of distances in multi-dimensional space occur in modern taxonomy. Classical taxonomy was based as much as possible on descriptions of certain morphological features, or characters, of organisms, and the features selected were those which did not show much variation within a species or other taxonomic unit (the taxon), but did show differences between taxa. In the higher plants, floral morphology has been of particular significance. Now, however, many other morphological characters (particularly those that are measurable) are used, and special methods are available to cope with the variability that exists. Measurements such as leaf length, leaf width and

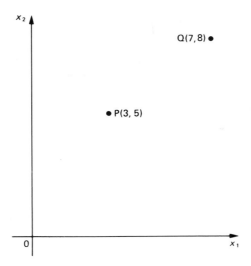

Fig. 3.7 The projection of two points situated in five-dimensional space on to the x_1–x_2 axial plane (see *example 3.2*).

internode length are made on a number of plants within a taxon, and if we consider just these three characters then the measurements made can be plotted on a three-dimensional diagram (Fig. 3.8). Because of the variability from plant to plant, a cluster of points would result; but if the mean value of the

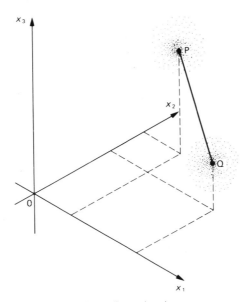

Fig. 3.8 Two clusters of points in three-dimensional space.

measurements of each character were calculated, these three mean values would be the co-ordinates of a single point P, in the centre of the cluster. Now, take a different but related taxon and repeat the process. Another cluster of points, together with a mean point Q, would result. Evidently if the two taxa are closely related, the distance D between them would be small, and *vice versa*. The distance will be given by

$$D = \sqrt{\{(X_1' - X_1)^2 + (X_2' - X_2)^2 + (X_3' - X_3)^2\}}$$

The above example was based on the measurement of three characters per plant, thus requiring a space of three dimensions which can be represented on a diagram. There is no reason why a greater number of characters should not be measured on each plant. Each character 'requires' a dimension in space, and the distance between two taxa is calculated by the 'D-formula', using as many terms as required on the right-hand side – one term for each character measured.

If interest centres on the taxonomic relationships within a critical genus such as *Rubus*, in which the *R. fruticosus* aggregate consists of a large number of microspecies, the method of finding distances between these taxa will help to shed light on their interrelationships and will therefore be of considerable use in classifying them. In fact, we can go further. The detailed methodology, based on the above principles, for dealing with the primary data can even attempt to answer the question of whether a particular taxon should be regarded as a full microspecies or merely as a variety of another microspecies.

Example 3.3

A certain genus of flowering plants contains three species, and we wish to enquire about their interrelationships. Four characters are measured on each of a number of individuals from each species, and the average for each character and species is calculated. The four characters measured, and the axis in space with which each is identified, are as follows:

x_1 – length of 3rd leaf from base of stem
x_2 – length of 3rd internode
x_3 – corolla diameter
x_4 – number of flowers per inflorescence

The results may be tabulated, thus:

	x_1	x_2	x_3	x_4
Species 1	2	5	2	8
Species 2	3	6	5	2
Species 3	7	4	3	7

Consider the first two characters only: the three species are shown plotted against the axes x_1 and x_2 on Fig. 3.9(a). The distances between the species are:

Species 1 – Species 2

$$D^2 = (3 - 2)^2 + (6 - 5)^2 = 1 + 1 = 2$$
$$D = \sqrt{2} \simeq 1.414$$

Species 2 – Species 3

$$D^2 = (7 - 3)^2 + (4 - 6)^2 = (4)^2 + (-2)^2 = 16 + 4 = 20$$
$$D = \sqrt{20} \simeq 4.472$$

Species 3 – Species 1

$$D^2 = (7 - 2)^2 + (4 - 5)^2 = (5)^2 + (-1)^2 = 25 + 1 = 26$$
$$D = \sqrt{26} \simeq 5.099$$

On the basis of the two vegetative characters, from the graph and the above results, it would appear that species 1 and 2 are fairly similar while species 3 is rather different from both the other species. If now we consider the two floral characters alone, we find the situation depicted in Fig. 3.9(b). Try the distance calculations yourself, but it is clear from the graph that, on the basis of floral characters, species 1 and 3 are similar while species 2 is the outlier. Finally, let us use all the information, and calculate the inter-species distance in the four-dimensional space required to represent the total information supplied. First, note the co-ordinates of the three species in four-dimensional space:

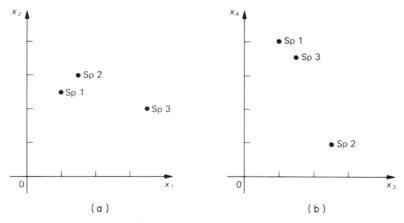

(a)　　　　　　　　　　　　　　(b)

Fig. 3.9 The projection of three points situated in four dimensional space on to: (**a**) the x_1–x_2 axial plane, and (**b**) the x_3–x_4 axial plane.

Species 1(2, 5, 2, 8), Species 2(3, 6, 5, 2), Species 3(7, 4, 3, 7). The distances are:

Species 1 – Species 2

$$D^2 = (3 - 2)^2 + (6 + 5)^2 + (5 - 2)^2 + (2 - 8)^2 = 47$$
$$D = \sqrt{47} \simeq 6.856$$

Species 2 – Species 3

$$D^2 = (7 - 3)^2 + (4 - 6)^2 + (3 - 5)^2 + (7 - 2)^2 = 49$$
$$D = \sqrt{49} = 7.000$$

Species 3 – Species 1

$$D^2 = (7 - 2)^2 + (4 - 5)^2 + (3 - 2)^2 + (7 - 8)^2 = 28$$
$$D = \sqrt{28} \simeq 5.292$$

Now it appears that the inter-species distances are more uniform, but again, species 1 and 3 are more similar to one another than to species 2. This confirms the finding when only the two inflorescence characters were used. The plant taxonomist may wish to place more emphasis on distances involving floral characters alone, which would be in keeping with the classical approach. This, however, is for the taxonomist to decide; mathematical methods can only provide the information, but the final evaluation is firmly the province of the biologist.

Linear functions

So far, we have only considered isolated points in space, how their positions may be defined, and the distances separating them measured. These concepts, although fundamental, are limited. To extend the usefulness of mathematics we need to be able to describe and analyse 'shapes' in space.

To fix ideas, let us consider these shapes firstly in two-dimensional, and then in three-dimensional, space. In two dimensions, say on the plane of a sheet of paper, the shapes referred to above would be curved lines, or simply, *curves*. Some curves enclose an *area* – an ellipse is one such example; but most regular curves do not enclose an area within themselves.

The three-dimensional analogue of a curve is a *surface*. Some surfaces, such as an ellipsoid (e.g. the shape of a rugby football), may enclose a *volume*. For all higher dimensions the corresponding terms are *hyper-surface* and *hyper-volume*, respectively. As in the previous sections of this chapter, we shall consider the second, third, and higher dimensions separately.

Two dimensions – the straight line

In two dimensions we are working with curves. From the mathematical point of view, a straight line is a special, and indeed the simplest, type of 'curve'; and the remainder of this chapter is devoted to a study of the straight line in two-dimensional space, and of its analogues in spaces of higher dimensions.

Euclidean geometry accurately describes a straight line through two definitions: (*i*) a line has length but no breadth, and (*ii*) a straight line is the shortest line joining two points. However, the usefulness of these concepts, which define the nature of a straight line, is limited by the fact that no information on the position of the line is conveyed. If the actual location of the straight line as well as its properties can be defined, a much more complete and useful picture emerges.

The properties and position of any shape in space can be completely specified in one of two ways. One way would be to draw an accurate graph of the shape from detailed tables giving co-ordinates of the points that lie on the curve. This method of extracting information about a shape is cumbersome and time-consuming, and has severe limitations.

The second method of describing completely the properties and location of a shape does not rely on graphical presentation at all, although the latter may be a useful adjunct. The necessary information is contained in an ***equation*** involving the two variable quantities x and y. The general form of the equation determines the type of shape we are dealing with, and particular numerical values of the constants in the equation serve to identify a particular member of the type or class of shape. Evidently to describe a straight line, we require the ***equation of the straight line***, and this can be obtained by a few algebraic manipulations based on the Euclidean properties of the straight line (referred to above), after specifying the position of the line.

Equation of the straight line which is specified by slope and intercept

There are two ways in which the position of a straight line can be defined on a plane relative to co-ordinate axes. These are: either (*i*) by specifying the orientation of the line with respect to the axes and the co-ordinates of any one point on it, or (*ii*) by specifying the co-ordinates of any two points on the line. Let us employ the first method of defining the position of a line. The orientation is given by the angle that the line makes with the x-axis, θ, and the point on the line selected for specification is that at which the line intersects the y-axis (Fig. 3.10a). This point is called the ***intercept***, and its co-ordinates are $(0, a)$. The reason for choosing this particular point is that this specification leads directly to a form of equation known as the ***standard equation of the straight line***.

The angle θ is not used directly in specifying the orientation of the line, as this would not lead to an equation involving x and y. Instead, we use the tangent of this angle and give it the symbol b. Thus $b = \tan \theta$, and this quantity is known as the ***slope***, or ***gradient***, of the line.

In order to derive the equation of the line shown in Fig. 3.10(a) from the specification of its position, we need to choose *any* point on the line other than the point already selected (i.e. the intercept). The co-ordinates of this other point are quite unspecific, and the point itself may be designated as $P(x, y)$.

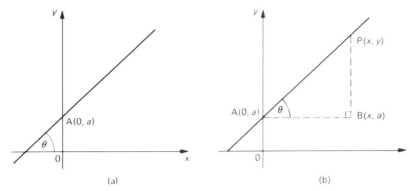

Fig. 3.10 (**a**) A straight line with gradient θ and intercept on the y-axis of a. (**b**) The same, but with construction lines added for determining the equation of the line.

This is shown in **Fig. 3.10(b)**, together with a constructed right-angled triangle PAB.

$$\text{Now} \qquad b = \tan \theta = \frac{\text{PB}}{\text{AB}} = \frac{y-a}{x-0}$$

Taking the purely algebraic parts from each end of the string of equalities, we have

$$b = \frac{y-a}{x}$$

Multiplying both sides by x gives us $\qquad bx = y - a$

and re-arrangement finally yields $\qquad y = a + bx \qquad\qquad (3.4)$

Equation 3.4 is the standard equation of the straight line. The two variables, x and y, are involved; also, there are two other quantities, a and b, which specify the position of the line relative to the x–y co-ordinate axes. For any one line, a and b are **constants** with numerical values, and then the equation enables us to specify the co-ordinates of any point on the line. This is done by first specifying the x-value of the point, and then using the equation to calculate the y-value. A particular value of a and of b defines a unique line; any other pair of values defines a different line. These constants (a and b) are sometimes called **parameters**.

Equation of the straight line which is defined by slope and any point

This is a more general case of the above. We are given a straight line of gradient b which passes through a particular point A(X, Y). The situation is

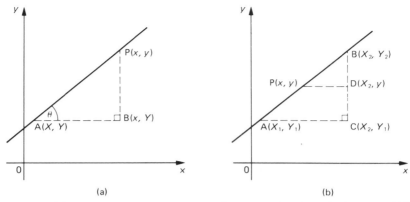

Fig. 3.11 (a) A straight line with gradient θ and passing through the point A(X,Y). (b) A straight line passing through the points A(X_1, Y_1) and B(X_2, Y_2).

depicted in Fig. 3.11(a): P(x, y) is any point on the line other than A, and a triangle is completed as shown with point B(x, Y).

$$\text{Now} \qquad b = \tan \theta = \frac{\text{PB}}{\text{AB}} = \frac{y - Y}{x - X}$$

$$\text{i.e.} \qquad y - Y = b(x - X)$$

$$\text{or} \qquad y = Y + b(x - X) \tag{3.5}$$

The form of the *equation 3.5* is quite useful in itself, as it shows explicitly the co-ordinates of the selected point on the line. However, the equation can easily be converted to the standard form, and this is

$$y = (Y - bX) + bx \tag{3.6}$$

and the quantity $(Y - bX)$ is the intercept (a in our previous notation).

Equation of the straight line which is defined by any two points

It is a fundamental fact of Euclidean geometry that any two points may be joined by a straight line. Suppose that we know the co-ordinates of two points, and are required to find the equation of the straight line joining them.

Let the two points be A(X_1, Y_1) and B(X_2, Y_2). Draw the line joining them, and let P(x, y) be any other point on the line. Complete the construction lines as shown in Fig. 3.11(b). Now the slope of the line is given by BD/DP if triangle BDP is considered, and by BC/AC if triangle ABC is considered.

Therefore, since the gradient of a straight line is the same however it is viewed, BD/DP = BC/AC. Thus

$$\frac{Y_2 - y}{X_2 - x} = \frac{Y_2 - Y_1}{X_2 - X_1}$$

After a few algebraic manipulations, this equation emerges in standard form, as follows:

$$y = \frac{X_2\,Y_1 - X_1\,Y_2}{X_2 - X_1} + \frac{(Y_2 - Y_1)}{(X_2 - X_1)}\,x \qquad (3.7)$$

Hence the slope of the line is $(Y_2 - Y_1)/(X_2 - X_1)$, which is obvious by inspection of Fig. 3.11(b), and the intercept is $(X_2\,Y_1 - X_1\,Y_2)/(X_2 - X_1)$.

The straight line with a negative slope

In deriving the various equations of the straight line, the diagrams have always shown the line *ascending* from left to right. The angle θ that the line makes with the x-axis, as shown in Figs. 3.10 and 3.11, lies between 0° and 90° thus yielding a positive value of tan θ. If we have a straight line which *descends* from left to right, the angle θ lies between 90° and 180° (Fig. 3.12). Now when 90° $< \theta <$ 180°, tan $\theta = -$tan (180° $- \theta$) (see *A Biologist's Advanced Mathematics*); thus for lines descending from left to right, tan θ, which is b, is a negative quantity. The standard equation of the straight line is still the same as *equation 3.4*, but b is now negative instead of positive.

The whole situation regarding the gradient of a straight line may now be summarized. When the line is horizontal, the gradient is zero. As the line moves away from the horizontal towards the vertical, sloping up to the right, the gradient increases through positive values. When the line is vertical, the

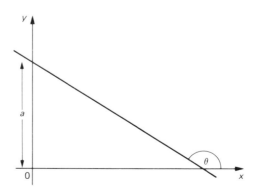

Fig. 3.12   A straight line with negative gradient.

gradient is infinitely large; and as the line moves away from the vertical towards the horizontal, sloping down to the right, the gradient decreases through negative values.

Example 3.4

Find the equations of the following straight lines:

(*a*) slope of 1.4 and passing through the point (0, −3.7);
(*b*) slope of −1, and passing through the point (−2, 3);
(*c*) passing through the points (3, 2) and (−2, −3).

(*a*) Since the point (0, −3.7) is on the *y*-axis, the intercept is −3.7. The equation can thus be written down directly:

$$y = -3.7 + 1.4x$$

(*b*) The slope is given as −1, and the line passes through the point (−2, 3). The appropriate form of equation is given by *3.5*, which in the present example becomes

$$y = 3 - (x + 2)$$

In standard form, this is $\qquad y = 1 - x$

(*c*) First, find the gradient of the line. This is given by the difference of the ordinates of each point divided by the difference between the abscissae:

$$b = \frac{2 - (-3)}{3 - (-2)} = \frac{5}{5} = 1$$

To obtain the intercept, one can either substitute directly into *equation 3.7*, or the co-ordinates of one of the two points can be substituted into the standard equation. Using the first point (3, 2), we have

$$2 = a + (1)(3)$$

giving $a = -1$. Hence, the equation in standard form is

$$y = -1 + x$$

Three dimensions – the plane

The equation of a straight line is a relationship between two mathematical variables, *x* and *y*. The two variables can be represented on a graph as co-ordinate axes, and the equation is depicted as a straight line whose position in relation to the axes depends on the values of the constants in the equation.

The *x*-variable is often called *independent*, and the *y*-variable *dependent*; when we use the equation of a straight line in standard form, we are often interested in seeing what value *y* assumes when a particular *x* is selected. The idea is that *y* depends on *x* according to the equation.

When considering three dimensions, there are three variables (axes), and mathematical relationships between them in the form of equations can be visualized and constructed. The 'shape' which the equation represents in three-dimensional space is a surface, which may be plane or curved. The equation itself may be considered to have two independent variables (axes), which can be designated as x_1 and x_2, and one dependent variable that can still be denoted as *y*.

In three dimensions, the analogue of the straight line is a plane surface, as shown in Fig. 3.13. When considering what form the equation to such a surface should take, we note that the plane intersects the *y*-axis; hence an intercept, *a*, can still be recognized. Next, consider a plane formed by the x_1 and *y* axes: the x_1–*y* axial plane. It can be seen that the plane surface crosses this axial plane as a straight line (Fig. 3.14a). This straight line makes an angle θ_1 with the x_1-axis, and so we may write $b_1 = \tan \theta_1$. A similar situation is found on the axial plane x_2–*y* (Fig. 3.14b): the straight line, in which the plane surface intersects the axial plane, makes an angle θ_2 with the x_2-axis, and so $b_2 = \tan \theta_2$. By a method analogous to that used for deriving the standard

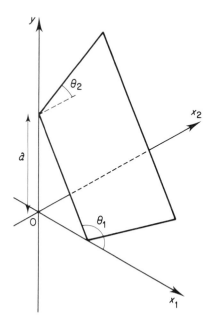

Fig. 3.13   A plane surface with gradient θ_1 on the x_1–*y* axial plane, gradient on the x_2–*y* axial plane, and intercept on the *y*-axis of *a*.

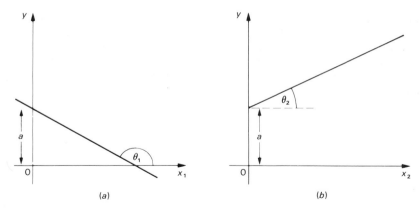

Fig. 3.14 The straight lines designating the intersections of the plane in Fig. 3.13 with: (**a**) the x_1–y axial plane, and (**b**) the x_2–y axial plane.

equation of the straight line, it can be shown that the equation of the plane surface is

$$y = a + b_1 x_1 + b_2 x_2 \qquad (3.8)$$

You can see that this equation is very similar to the standard equation of the straight line, but because we now have two independent variables instead of one, an extra term of the form bx is added.

The straight lines shown in Fig. 3.14, which are the intersections of the plane surface with the axial planes x_1–y and x_2–y have the equations $y = a + b_1 x_1$ and $y = a + b_2 x_2$, respectively.

Example 3.5

Find the equations of the following planes:

(*a*) gradient of 1.7 in one direction, gradient of -2.1 in the direction at right-angles to the previous direction, and passing through the point $(0, 0, 1)$;

(*b*) gradients of 0.5 and 1, and passing through the points $(2, -3, 1)$;

(*c*) having a gradient of 3 in one direction, and passing through the points $(0, 1, 1)$ and $(1, 3, 2)$.

(*a*) The order of the axes represented in the co-ordinate specification is, by convention, (x_1, x_2, y). Hence the point $(0, 0, 1)$ is the intercept on the y-axis ($=1$), and so the equation can be written down directly:

$$y = 1 + 1.7x_1 - 2.1x_2$$

(*b*) Since we know the two gradients, we can first write the equation as $y = a + 0.5x_1 + x_2$. Further, we know that the point $(2, -3, 1)$ lies on the surface; so, substituting into the preliminary equation, we have

$$1 = a + (0.5)(2) + (-3)$$

i.e. $1 = a + 1 - 3$ giving $a = 3$

So the final equation is $y = 3 + 0.5x_1 + x_2$

(*c*) Since the information given about the gradient is not specific about the direction, we can write two possible preliminary equations:

$$y = a + 3x_1 + b_2 x_2$$

or $y = a + b_1 x_1 + 3x_2$

We are also given the co-ordinates of two points that lie on the surface; so using one of the above preliminary equations, we can form two equations in two unknowns – the intercept, a, and one of the gradients. Using the first of the preliminary equations, we have

$$1 = a + 3(0) + b_2(1)$$

and $2 = a + 3(1) + b_2(3)$

and solving this pair of equations gives $a = 2$ and $b_2 = -1$, giving the equation of the plane surface as

$$y = 2 + 3x_1 - x_2$$

Substituting the co-ordinates of the two points successively into the second preliminary equation gives

$$1 = a + b_1(0) + 3(1)$$

and $2 = a + b_1(1) + 3(3)$

which yield the solution $a = -2$ and $b_1 = -5$. So now the equation of the plane surface is

$$y = -2 - 5x_1 + 3x_2$$

If three points had been specified, and no information about any gradient given, it would have been necessary to solve three simultaneous equations in three unknowns, unless one of the given points was the intercept when we should have two equations in two unknowns (see *A Biologist's Advanced Mathematics*).

n dimensions – the hyperplane

As usual, more than three dimensions cannot be illustrated; but, as we saw when specifying the position of a point, and calculating distances between points, formulae established for two and three dimensions extend quite easily to higher dimensions.

In the present situation we have n dimensions, which means $(n-1)$ independent variables $(x_1, \ldots, x_{(n-1)})$ and one dependent variable (y). The analogous equation to the two-dimensional straight line (*3.4*) and the three-dimensional plane surface (*3.8*) is

$$y = a + b_1 x_1 + b_2 x_2 + b_3 x_3 + \ldots + b_{(n-1)} x_{(n-1)} \qquad (3.9)$$

Any axial plane formed by the y-axis and any one x-axis, x_i, will have an equation of the form

$$y = a + b_i x_i \qquad i = 1, \ldots, (n-1) \qquad (3.10)$$

Expression 3.10 means that i can have any value from 1 up to $(n-1)$ inclusive. For instance, if $i = 5$, *equation 3.10* becomes $y = a + b_5 x_5$ and represents the straight line given by the intersection of the hyperplane with the axial plane x_5–y.

Linear functions

The most general equation that we have encountered in this section of the chapter is *equation 3.9*; the other similar *equations*, *3.4* and *3.8*, are simply sub-sets of the right-hand side of *3.9*. The two most important features of *equation 3.9* are: (*i*) that all the variables, the x_i and y, are of degree one (i.e. there is no term such as y^2 or x_4^3); and (*ii*) that only one x_i or y appears in any term (i.e. there is no term of the form $x_i y$, or $x_i x_j$ $i \neq j$). This kind of equation is known as a **linear** equation, because the simplest equation of this type is that of a straight line.

Therefore, a linear equation in two variables is the equation of a straight line. A linear equation in three variables is the equation of a plane, while a linear equation in four or more variables is the equation of a hyperplane.

Instead of the word 'equation', the term 'function' (short for 'mathematical function') is more usual. Thus, in the equation of a straight line, $y = a + bx$, we say that y is a linear function of x. In the equation of a plane, $y = a + b_1 x_1 + b_2 x_2$, we say that y is a linear function of x_1 and x_2; and so on. In describing the relationship this way, we again emphasize the dependent nature of the variable y, making it a function of one or more x_i which are given.

There are, of course, many types of mathematical functions other than linear ones, and we shall meet some useful ones in the next and later chapters.

Linear functions in biology

Choosing and fitting a function

The use of a mathematical function in biology is to provide a quantitative description of the way some biological activity proceeds in relation to some factor external to the biological system. Some examples are: the rate of photosynthesis in relation to light intensity, the growth of a population with respect to time, the change in size of some character of a species of organism in relation to altitude above sea level. In these examples, the x-axis would represent the external factor – light intensity, time, altitude; and the y-axis represents the biological **response** – photosynthetic rate, number of organisms in a population, size of a particular character in a certain species.

In examples of this kind, one is interested in a possible mathematical relationship (in the form of a function) between the two variables, to describe the biological response to changes in the external factor. This raises two questions: (*i*) what kind of mathematical function is involved and, when this question has been answered, (*ii*) what are the values of the constants, or parameters, of the function? In some instances, which are rare in biology, theoretical considerations can provide the answer to the first question. More commonly, however, one must find a function which reasonably describes the situation by carrying out a suitable experiment, plotting the observations on a graph, and examining for trends. A function selected in this way gives an **empirical**, rather than a theoretical, description of the biological process.

From the remarks made towards the end of Chapter 1 (page 5), it is obvious that the experimental observations will not lie *exactly* on the curve of any function, and so the choice of function becomes more difficult. In many situations, however, a linear function provides an adequate empirical description of the response of a biological process to variations in level of an external factor. It is considerably easier to detect by eye deviations from a straight line than deviations from any other type of curve, and so linear functions find considerable use in biology.

As an example, consider an experiment to assess the way in which the uptake of an ion by plant tissue is affected by temperature. The experiment would consist of immersing samples of the tissue in a series of vessels holding a solution containing the ion, each vessel being maintained at a different temperature. After a certain fixed interval of time the plant tissue is fixed and assayed for the content of the ion, and the results (amount of the ion taken up at each temperature) are plotted on a graph, as in Fig. 3.15(a). Plotted in this way, the data suggest a straight line.

When the type of function has been decided upon, the constants of the function have to be estimated for the particular set of data in hand. For our present example, we have selected a linear function as suitable, and we now require values for the constants a and b of a straight line that will best describe the data shown in Fig. 3.15(a). The process of estimating the constants of a function is termed *fitting a function*, in general; and in this example, where the function to be fitted is linear, the process is usually described as *fitting a straight line* to the data.

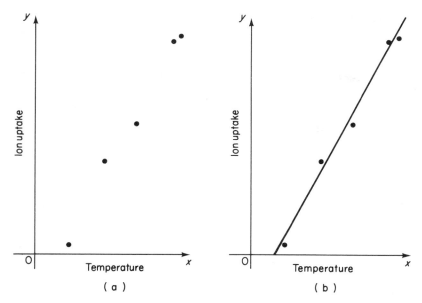

Fig. 3.15 (**a**) An hypothetical set of data from an experiment on the response of uptake of an ion by plant tissue at different temperatures. (**b**) The same, but with a 'fitted' straight line (see page 40).

The fitting process requires a knowledge of statistics, and so is outside the scope of this book; but let us assume that a straight line is now fitted to the data of Fig. 3.15(a), and that the line is drawn on the graph by finding the co-ordinates of two points on the line and joining them up. The final graph now appears as in Fig. 3.15(b).

The uses of a fitted function

Broadly speaking, there are three uses of a fitted mathematical function: (*i*) smoothing of the data, (*ii*) prediction purposes, and (*iii*) use of the mathematical function itself (i.e. its form and parameter values) to provide an insight into the working of the process under study. The only use of (*i*) is to provide a clearer picture of the trend involved when presenting the results graphically.

Prediction means specifying a value of *x*, and asking what the corresponding value of *y* is likely to be. In the present example, one can enquire about the most likely quantity of ion taken up at a given temperature. A word of warning is necessary, however, as one can either make predictions within the range of the data (***interpolation***) or outside the range of the data (***extrapolation***). The range of the data in the example of ion uptake extends from the lowest temperature employed *in the original experiment* to the highest temperature used. It is safe to interpolate, but not to extrapolate. The reason is that a linear function has been fitted to actual experimental data on the basis that, over the range of temperatures covered by the experiment, such a function appears to provide a reasonable description of the situation. It is certainly not safe to assume that this linear function will adequately describe the situation at lower or higher temperatures than those embraced by the experiment. Indeed, common sense tells us that at very high temperatures the whole biological system breaks down, and could certainly no longer continue in a linear mode.

If one has empirically fitted a function to a set of experimental data, then the use described in (*iii*) above should be employed with great caution. If one has theoretically deduced the existence of a particular function, before fitting the function to experimental data, then the constants can be used, perhaps (in the ion uptake example) to compare the uptake of different ions by the same plant species, or to compare the uptake of the same ion by different species. Even when the function is an empirical description of the data, comparisons of this kind may be made, so long as nothing more fundamental is attributed to the parameters of the equation; for in this case the parameter values found will simply be characteristic of the particular experiment that was carried out.

Extension to more than one external factor

The above type of analysis can be extended to cover more than one external factor, and then a space of more than two dimensions will be required. In the ion uptake example, we might be interested in the effects of both temperature and pH; so these two factors would be identified with axes x_1 and x_2 of a three-dimensional graph, while the uptake of ion remains as the y-axis. If it is assumed that a linear relationship also holds over a restricted range of pH with regard to ion uptake, the entire situation can be described by *equation 3.8*. In theory, there is clearly no limit to the number of external factors that can be included in a single investigation; but in practice it may be impossible to include very many external factors in the same investigation because of the large size of the experiment needed, and the subsequent difficulty of interpreting the results.

EXERCISES

1. Find the distances between the pairs of points whose co-ordinates are:

    (*a*) (−1, 3) and (2, −4)
    (*b*) (0, 3, 4) and (−2, −1, 5)
    (*c*) (1, 1, −1, 2) and (−2, −3, 4, 4)

2. The following table gives details of various morphological measurements made on common members of the genus *Veronica* (Speedwell) in Britain. All measurements are in mm, and each figure is the arithmetic mean of a number of individuals.

Species	Leaf length	Leaf width	Petiole length	Internode length	Pedicel length	Corolla diameter
V. beccabunga	35.00	17.58	3.42	68.00	29.92	6.83
V. officinalis	14.80	9.40	3.10	20.80	19.90	5.20
V. montana	22.19	17.56	9.25	38.25	38.63	8.94
V. chamaedrys	16.30	14.46	1.06	29.12	26.21	11.39
V. serpyllifolia	5.40	2.84	0.44	3.56	2.56	4.84
V. hederifolia	8.58	9.70	2.67	7.58	5.61	5.21
V. persica	11.13	10.70	1.83	6.83	17.00	9.87
V. filiformis	5.49	6.85	1.97	6.82	19.52	11.67

Draw graphs to show the positions of the species with respect to axes representing various pairs of characters. Also, calculate the distance between the species, as characterized by the above figures, in the full six-dimensional space. Compare your answers with the taxonomic arrangement of the species in the genus *Veronica* in a standard Flora.

3. Find the equations of the following straight lines:

   (a) making an angle of 60° with the x-axis, and passing through the point (2, 1.3495);

   (b) passing through the points $(-1, 0)$ and $(5, 4)$.

4

Functions and curves

In the previous chapter, we discussed the idea that a straight line in a two-dimensional plane could be completely described by an equation, $y = a + bx$, where x and y are the co-ordinate axes. For various reasons such an equation is best called a mathematical function; in this instance, y is a function of x. The equation also contains two constants, a and b, and any particular pair of values of these constants defines a unique straight line. When we come to examine the *form* of the straight line function, we see that it consists of three terms: one of the terms contains the variable x, another the variable y and the third term contains neither x nor y but is merely a constant. Moreover, there is no term containing x and y together, neither are there any terms in which x and/or y are raised to a power (other than to the power 1, of course). All these properties serve to define the form of the equation of a straight line, and so we know that any equation with these properties will be that of a straight line. For example, the equation $x - 2y + 3 = 0$ represents a straight line because its form fulfils all the requirements listed above. This equation can be rearranged into a more familiar form: $y = \frac{1}{2}x + \frac{3}{2}$; which shows that the gradient of the line is $\frac{1}{2}$ and that the intercept is $1\frac{1}{2}$.

A mathematical function whose form differs in *any* way from the specification contained in the previous paragraph will not represent a straight line, but a **curve**. Any particular form of function gives rise to a unique type of curve, and the number of different functional forms and their corresponding curves is indefinitely large. Fortunately, however, many types of function and their curves can be grouped into 'families' and, as biologists, we need only concern ourselves with a few of these families. Two families of function will be introduced in this chapter, and others will appear later in the book.

Polynomial functions

The standard equation of a straight line is $y = a + bx$: on the right-hand side there is a term consisting of a constant only, and a term consisting of the variable x multiplied by a constant. A constant which multiplies a variable is called a **coefficient**. If another term of similar form is added, consisting of x^2 and a coefficient, the above equation becomes $y = a + bx + cx^2$. Other terms could be added to the right-hand side, in which the power of x in increased by one for each new term added. Thus we have $y = a + bx + cx^2 + dx^3$ and $y = a + bx + cx^2 + dx^3 + ex^4$, for example, where a, b, c, d, and e are constants, and the number of equations that can be set up by this process is clearly infinite. Evidently, all the functions of this type can be regarded as comprising a family, and collectively they are known as ***polynomial functions***, or just ***polynomials***.

Nomenclature

The names applied to the first few members of the polynomial family are summarized in the table below.

Equation	Polynomial name	Special name
$y = a + bx$	first degree polynomial	linear function
$y = a + bx + cx^2$	second degree polynomial	quadratic function
$y = a + bx + cx^2 + dx^3$	third degree polynomial	cubic function
$y = a + bx + cx^2 + dx^3 + ex^4$	fourth degree polynomial	quartic function
$y = a + bx + cx^2 + dx^3 + ex^4 + fx^5$	fifth degree polynomial	quintic function

Second degree polynomial

Curves of the equation $y = a + bx + cx^2$ are shown in Fig. 4.1. The actual position of the curve relative to the co-ordinate axes will, of course, depend on the values of a, b, and c. But the orientation of the curve, that is, whether the open end faces upwards or downwards, depends only upon the sign of c, the

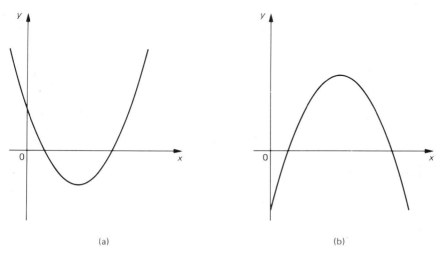

(a)   (b)

Fig. 4.1   Curves of the second degree polynomial function, $y = a + bx + cx^2$ with real roots: (**a**) c is positive, and (**b**) c is negative.

coefficient of x^2; in other words, upon the coefficient of the term containing the highest degree of x in the equation.

In a quadratic function wherein c is positive, working from left to right along the x-axis, the curve descends to a minimum value of y and subsequently rises again (Fig. 4.1a). The curve is symmetrical about a vertical line drawn through the lowest point. A curve in which c is negative rises to a maximum value and then descends (Fig. 4b); it has the same kind of symmetry as the curve with c positive.

The roots of the second degree polynomial

At the points where the curve intersects the x-axis, $y = 0$, which gives the equation

$$a + bx + cx^2 = 0 \qquad (4.1)$$

the (once!) familiar quadratic equation of school-days. Solving this equation may yield two values of x, which specify the points of intersection of the curve with the x-axis. These x-values are called the **roots** of the *equation 4.1*, and they can be evaluated using the formula

$$x = \frac{-b \pm \sqrt{(b^2 - 4ac)}}{2c} \qquad (4.2)$$

with the coefficients as defined in *equation 4.1*.

A quadratic curve does not necessarily intersect the x-axis. In *equation 4.2*, the quantity within the square root sign, $(b^2 - 4ac)$, is known as the **discriminant** of *equation 4.1*. If $b^2 > 4ac$, then the discriminant is positive and has a real square root. Thus the two points of intersection of the curve and the x-axis are at $x = \{-b + \sqrt{(b^2 - 4ac)}\}/2c$ and at $x = \{-b - \sqrt{(b^2 - 4ac)}\}/2c$, and the situation is as shown in Fig. 4.1. If, however, $b^2 < 4ac$, the discriminant is negative and the square root is a complex number (page 8). Under these conditions the quadratic equation has no real roots, and the curve does not intersect the x-axis (Fig. 4.2). The intermediate situation is when $b^2 = 4ac$, giving a discriminant of zero. Now the curve just touches the x-axis at one point only, given by $x = -b/2c$.

Example 4.1

Find the roots, if real, of the following equations

$$
\begin{aligned}
&(a) && 2x^2 - 7x + 4 = 0 \\
&(b) && 5x^2 - 4x + 2 = 0 \\
&(c) && x^2 - 6x + 9 = 0
\end{aligned}
$$

(*a*) The left-hand side of the equation is the opposite way round from that quoted in *equation 4.1*, so be careful when matching the coefficients; it is the

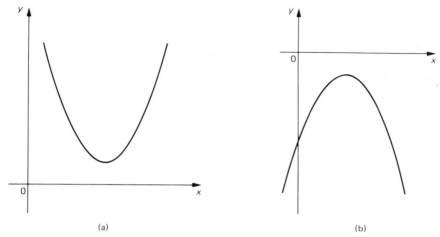

(a)   (b)

Fig. 4.2   Curves of the second degree polynomial function $y = a + bx + cx^2$ with non-real roots: (**a**) c is positive, and (**b**) c is negative.

powers of x in the various terms that matter, and not the order of appearance of the terms from left to right. So $a = 4$, $b = -7$, $c = 2$.

$$\text{Then} \quad x = \frac{7 \pm \sqrt{(49 - 32)}}{4} = \frac{7 \pm \sqrt{17}}{4} = \frac{7 \pm 4.123}{4}$$

$$\text{So either} \quad x = \frac{7 + 4.123}{4} = \frac{11.123}{4} = 2.78$$

$$\text{or} \quad x = \frac{7 - 4.123}{4} = \frac{2.877}{4} = 0.72 \quad \text{(correct to 2 decimal places)}$$

(*b*) Here, we have $a = 2$, $b = -4$, $c = 5$.

$$\text{So} \quad x = \frac{4 \pm \sqrt{(16 - 40)}}{10}$$

The discriminant is negative, and so there are no real roots.

(*c*) In this example, $a = 9$, $b = -6$, $c = 1$.

$$\text{So} \quad x = \frac{6 \pm \sqrt{(36 - 36)}}{2} = 3$$

and the equation has only one real root or, as is sometimes said, two coincident roots.

Third degree polynomial

Curves of the function $y = a + bx + cx^2 + dx^3$ are shown in Fig. 4.3. As in the case of the quadratic curve, the actual position of a cubic curve relative to the co-ordinate axes depends on the values of the four constants: a, b, c, and d. There are two different orientations (Fig. 4.3(a and b) and (c and d)) according

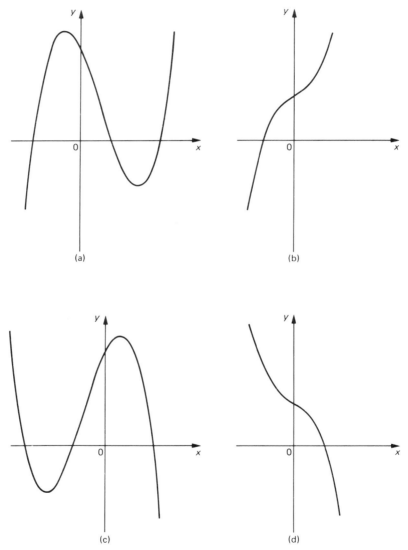

Fig. 4.3   Curves of the third degree polynomial function $y = a + bx + cx^2 + dx^3$: (**a**) and (**b**) d is positive, (**c**) and (**d**) d is negative.

to the sign of d, the coefficient of x^3 (i.e. the coefficient of the term containing the highest power of x in the equation). Moreover, there may or may not be a loop in the curve (Fig. 4.3(a and c) and (b and d)), again dependent on the values of the coefficients.

The roots of a third degree polynomial are very much more difficult to determine than those of a quadratic function, and discussion of this topic appears in *A Biologist's Advanced Mathematics*.

Polynomials in biology

Polynomial functions and their curves are very important in mathematics, as they have relatively simple mathematical properties. Accordingly, we shall make considerable use of polynomials in succeeding chapters to illustrate ideas and methods. On the other hand, in the mathematical description of biological phenomena polynomials find little place; but they may be useful for smoothing and forecasting purposes.

The power function

Each term in a polynomial function is of the form ax^n; even the first term, a, can be written in this form, i.e. ax^0, since $x^0 = 1$ (page 10). Thus in each term of a polynomial function, a is any real number and n is a positive integer (or whole number) including zero.

Now, consider a function wherein there is only *one* term of the form ax^n, but with the properties of a and n almost reversed; that is, n is now *any* real number and a is *any positive* real number. So the full specification of the relationship between x and y is

$$y = ax^n \qquad a, x > 0 \qquad -\infty < n < \infty \qquad (4.3)$$

This relationship is called the power function, because x is raised to a *power*, n (page 9).

If, in a particular power function, n is an integer, it is easy to calculate y for a given x and so construct a graph of the function; but if n is not an integer, then logarithms have to be used in the way described at the end of Chapter 2 (page 15). Taking logarithms (to any convenient base, although as we shall see later there is good reason to use natural logarithms) of both sides of *equation 4.3*, we obtain

$$\log_e y = \log_e (ax^n)$$

By using the result of Theorem 2.4 on page 12, the right-hand side can be recast:

$$\log_e y = \log_e a + \log_e x^n$$

and finally, applying the result of Theorem 2.6, we have

$$\log_e y = \log_e a + n \log_e x \qquad (4.4)$$

Equation 4.4, which has been derived from the original *function 4.3* merely by taking logarithms of both sides, is particularly interesting. First, notice that the variable x is no longer raised to a power; indeed, x has been replaced by $\log_e x$ along with change of y to $\log_e y$. Secondly, *equation 4.4* is linear in form between the two variables $\log_e x$ and $\log_e y$. This means that if we plot $\log_e y$ against $\log_e x$ for a particular power function, we shall obtain a straight line of gradient n and intercept $\log_e a$ (Fig. 4.4(c and d)). If n is positive, the line slopes up to the right, and *vice versa*.

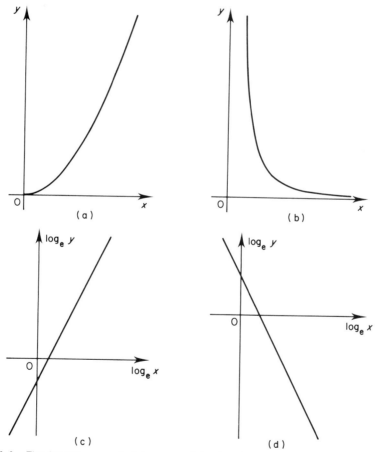

Fig. 4.4   The function $y = ax^n$: (**a**)   curve of the function when $n > 0$, (**b**)   curve of the function when $n < 0$, (**c**) the straight line $\log y = \log a + n \log x$ corresponding to (**a**), (**d**) the straight line corresponding to (**b**).

To obtain the graph of the original function, $y = ax^n$, we need add only one extra step to those outlined in the previous paragraph, namely to take the antilogarithms (natural, e^x) of each of the pairs of values ($\log_e x$, $\log_e y$). This gives corresponding x- and y-values, enabling the graph of y against x to be plotted.

The process of changing a variable into another variable by means of a definite mathematical rule is known as a *transformation*; in the present instance we have applied a *logarithmic transformation* to both x and y. Simple transformations, such as the logarithmic, are very useful in both bio-mathematics and biostatistical methods, and we shall meet other transformations in this book.

In the graphs of the transformed variables (Fig. 4.4(c and d)) the straight line can exist in any of the four quadrants formed by the co-ordinate axes, but owing to the restrictions placed on the values of a and x (both positive) y is positive too, and so the curves of y against x can exist only in the upper right-hand quadrant. For positive n ($n > 0$) the curve starts at the origin, whatever the value of a, and bends up to the right (Fig. 4.4a). For $n < 0$, however, the curve is quite different: it bends down to the right, but however large x becomes, y never becomes zero or negative; that is, the curve does not meet or intersect the x-axis. Conversely, at the top left-hand end, the curve approaches but does not intersect the y-axis. This kind of behaviour is an extremely important mathematical phenomenon, and it is very relevant in the mathematical description of biological processes: it is a topic that will recur in greater detail later in this chapter, and at intervals throughout the book.

Allometry

Very many years ago, zoologists found that when length measurements made on two different body parts (x and y) of growing individuals of an animal species were plotted in the form $\log_e y$ against $\log_e x$ a linear band of points appeared on the graph. Evidently, the two length measurements conformed approximately to the power function

$$\log_e y = \log_e a + b \log_e x$$

Subsequently, this relationship was found to have much wider biological validity – length measurements of plant parts, weight measurements of plant parts, and even between respiration rates and body size in animals. The name 'allometry' has long been employed to describe this kind of relationship: either one can say that two measurements have an allometric relationship between them or, because such measurements arise as a result of growth, that the two parts are growing allometrically.

In plants, an allometric relationship between the dry weights of two parts may show an abrupt change of gradient, and thus can often be associated with a physiological change (see the example in Fig. 4.5). Indeed in plants, as will be

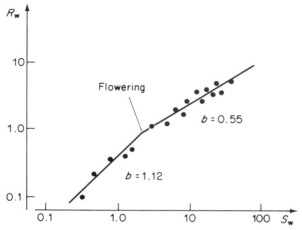

Fig. 4.5 A change in allometric relationship between roots and shoots of *Lolium multiflorum* (Italian rye-grass) at the onset of flowering (after Troughton, 1956). Notice that the two axes have logarithmic scales.

shown in Chapter 10 (page 180), the gradient of an allometric line has an important physiological significance.

Functional notation

In this chapter, two types of mathematical function have been examined. Let us take as an example for the present discussion the second degree polynomial:

$$y = a + bx + cx^2 \qquad (4.5)$$

Now the essence of this function lies on the right-hand side of the equals sign, since the quadratic function represented in *4.5* is a function *of x*. In fact, we invoke the other variable, *y*, only for convenience so that we may draw a graph of the function on the usual *x–y* axes. A way of writing *equation 4.5* without the use of *y* is

$$f(x) = a + bx + cx^2 \qquad (4.6)$$

and the left-hand side of *4.6* is read: 'function of *x*', or simply 'function-*x*'.

One important feature of a given mathematical function of *x* is that it can normally be evaluated numerically for any given value of *x*. Let us consider the function

$$f(x) = 3 + 2x - x^2$$

If $x = 0$ $f(0) = 3 + 2(0) - (0)^2 = 3$
    $x = 1$ $f(1) = 3 + 2(1) - (1)^2 = 4$
    $x = 5$ $f(5) = 3 + 2(5) - (5)^2 = -12$
    $x = -2$ $f(-2) = 3 + 2(-2) - (-2)^2 = -5$

There is another, equally important, use of the functional notation. In many situations, a mathematical function is known or inferred to exist between two variables, but the actual form of function may be in doubt or completely unknown. Under these circumstances, we may write

$$y = f(x) \qquad (4.7)$$

which implies the existence of a mathematical functional relationship between x and y, but does not specify its type. The letter f in the notation could be some other letter; thus $y = f(x)$, $y = F(x)$, $y = \phi(x)$ denote three different functional relationships between x and y. The notation can also be extended to cover the situation where y is a function of more than one variable. For example, we may write

$$y = f(x_1, x_2) \qquad (4.8)$$

which implies that y is a function of both x_1 and x_2 (cf. page 39).

Some properties of curves

The gradient of a curve

When discussing the straight line, we found that its orientation (slope or gradient) contributed towards its definition. The gradient of a straight line is the tangent of the angle the line makes with the horizontal axis, measured in an anti-clockwise direction; this quantity is given the symbol b and it appears as the coefficient of x in the equation describing the line.

A straight line is, however, a single special case of the more general phenomenon of curves. Can the idea of a gradient, which is an intuitively obvious property of a straight line, be extended so that it may be applied to a curve? In other words, can we speak of 'the gradient of a curve', and if so, how is it defined? The concept of the gradient of a curve is indeed valid, and is approached through two definitions.

(*i*) The **tangent to a curve** *at a particular point* is defined as that straight line which just touches, but does not intersect, the curve at the point under consideration.

(*ii*) The **gradient of a curve** *at a particular point* is defined as the gradient of the tangent of the curve at that point.

In Fig. 4.6(a) is shown a curve of some function $y = f(x)$ with tangents at two points, P and Q. It is impossible in a diagram to show that a straight line can touch a curve at one definite point only, but this is only because a drawn curve and line each have a finite thickness. The strict geometrical definitions of lines and curves state that they are infinitely thin. However, it is obvious that the tangent to the curve at point P is not the same as the tangent to the curve at point Q; the important distinction is that the *gradients* of the two tangents

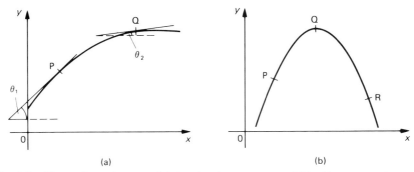

Fig. 4.6   The gradient of a curve: (**a**) showing the tangents, and (**b**) without tangents.

differ. The gradient of a straight line is the same wherever it is examined, but the gradient of a curve varies from point to point along the curve.

Although the gradient of a curve has been defined in terms of the gradient of a tangent to the curve, we can always think of the gradient of a curve without a direct reference to a tangent. In Fig. 4.6(b), it is evident that the gradient of the curve at point P is positive because the tangent at that point has a positive gradient; the gradient at point R is negative for a similar reason, and the gradient of the curve at point Q is zero because the tangent at that point is a horizontal line (gradient zero).

Hence, the concept of the gradient of a curve is a valid one, and for a particular curve whose equation is known it is possible to derive a formula which enables us to evaluate the gradient of the curve at any desired point. The gradient of a curve and its evaluation form the basis of the differential calculus.

Asymptotes

An *asymptote* is formally defined as a tangent to a curve at infinity. In more diagrammatic, but less accurate, terms it is a line towards which a curve may tend, but which the curve never intersects or meets. We have already encountered such a situation in the curve of the function $y = ax^n$. Consider the case where the constant n is negative, when the curve of the equation will be as shown in Fig. 4.4(b). With increasing values of x, y decreases but never becomes zero or negative. Hence on the graph as one moves further over to the right-hand side, the curve descends towards the x-axis but never actually touches or intersects it. Thus, the x-axis is an asymptote to the curve of $y = ax^n$ when $n < 0$.

It may be wondered just what significance such a concept as this could have in practical situations that a biologist might encounter. An example will be presented in the next section of this chapter, but one situation can be described in a general way here.

Frequently it is desirable to describe growth mathematically. This first necessitates measuring some attribute of growth (e.g. weight or length) of an

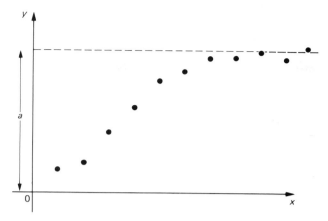

Fig. 4.7 Data requiring the fitting of a function with an asymptote $y = a$.

organism, or group of organisms, at several specified points in time, and plotting the results on a graph in which the x-axis represents time and the y-axis the growth attribute (Fig. 4.7). It is now necessary to select a function whose curve will pass as close to the observed points as possible. As discussed in Chapter 3, this requires us first to select a suitable function type and then to estimate the constants of this function from the experimental data. It is the first of these two problems, in relation to growth data, that we now discuss.

The growth of most organisms, particularly of animals, is characterized by the ultimate acquisition of a more or less constant maximum size which could, in theory, be maintained indefinitely were it not for the fact that senescence ensues. However, the phenomena of growth and senescence are not usually studied simultaneously, so that a physiologist in studying growth assumes that a maximum size is maintained constant. To describe this kind of situation mathematically, he would require a function whose curve is asymptotic to a horizontal line $y = a$ (Fig. 4.7) where a is the maximum size attained by the organism concerned. Some realistic 'growth functions' of this kind are described in Chapter 9.

Rectangular hyperbolae

The function $y = a/x$

It might be thought that a fairly complicated equation would be required to describe a curve with an asymptote. We can see that this is not the case by considering the function $y = a/x$, where a is a constant. Let us simplify further by putting $a = 1$, so that the function is now

$$y = \frac{1}{x} \qquad (4.9)$$

Consider what happens to y in *equation 4.9* as x increases in size: to be specific, let x take successively the values 1, 10, 100, 1000, etc., then the corresponding values of y will be 1, 1/10, 1/100, 1/1000, etc., i.e. 1, 0.1, 0.01, 0.001, etc. Evidently, as x increases indefinitely y will get smaller, and will approach, but never actually reach, zero; the line $y = 0$ (i.e. the x-axis) is an asymptote. The x-axis is, however, not the only asymptote; consider what happens to y when x gets very small. Let x take successively the values 1, 0.1, 0.01, 0.001, etc., then y will be correspondingly 1, 10, 100, 1000, etc. In other words, as y increases indefinitely x approaches zero; hence the line $x = 0$ (i.e. the y-axis) is also an asymptote. The curve of $y = 1/x$, resulting from our discussion so far, is shown in Fig. 4.8(a).

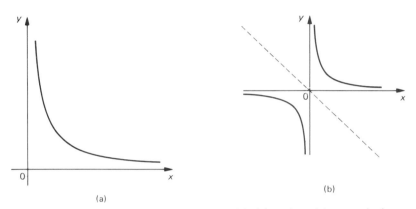

Fig. 4.8   Curves of the rectangular hyperbola $y = 1/x$: (**a**) portion of the curve in the upper right-hand quadrant defined by the co-ordinate axes, and (**b**) the complete curve.

Hitherto, we have considered only positive values of x in the equation $y = 1/x$, and these give rise to positive values of y. If x takes negative values, the corresponding values of y will likewise be negative. In fact, the complete curve of the function $y = 1/x$ is as shown in Fig. 4.8(b), and it is interesting to observe that the curve is in two distinct parts. If a straight line is drawn through the origin (broken line in Fig. 4.8b) with a gradient of -1, the two parts of the curve are mirror images of one another on either side of this line. The curve of the function $y = a/x$ has a similar appearance to the foregoing curve for any positive value of the constant a; the value of a merely places the curve closer to or further away from the x-axis for a given value of x. If, however, a were negative, the two parts of the curve would be in the other two quadrants formed by the co-ordinate axes.

The conclusion to be drawn from the above discussion is that the x- and y-axes are both asymptotes to the curve of $y = a/x$. Because the two asymptotes are at right-angles to one another, and because the function is a member of a family of functions known as ***hyperbolae***, the function $y = a/x$ is called a ***rectangular hyperbola***.

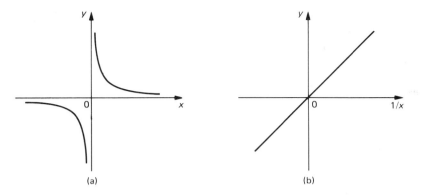

Fig. 4.9 (a) Curve of the rectangular hyperbola $y = 1/x$. (b) The straight line obtained when y is plotted against $1/x$.

The equation $y = a/x$ can be written in the form $y = a(1/x)$. Now if $1/x$ is plotted along the horizontal axis instead of x, then the relationship between y and $1/x$ is linear (Fig. 4.9(a and b)). This is a similar result to that encountered with the function $y = ax^n$, where it was found that plotting $\log y$ against $\log x$ gave a linear relationship. In the present case, the function $y = a(1/x)$ represents a straight line with gradient a and intercept zero.

The Michaelis–Menten function

In a biochemical reaction which is controlled by a single enzyme, the velocity, v, of the conversion of a single substance, for a fixed amount of enzyme, is given by

$$v = \frac{v_{max}\, s}{K_m + s} \qquad (4.10)$$

where s is the concentration of the substrate being converted, and v_{max} and K_m are constants. The subscripts of the constants make them look more complicated than if they were unsubscripted, but the difficulty is more apparent than real, and this notation is standard in biochemistry. The curve is shown in Fig. 4.10(a) and, although only the portion of the curve in the upper right-hand quadrant of the co-ordinate axes corresponds with real experimental data, part of the curve in the lower left-hand quadrant is shown as well, so as to indicate the curve's complete structure.

The constants have direct interpretations: v_{max} is the value of the horizontal asymptote – the maximum velocity approached when substrate concentration is indefinitely large; K_m is the substrate concentration at which the velocity is half the maximum. This latter parameter is possibly the more important one of the two, and is named the Michaelis constant. Under defined conditions of

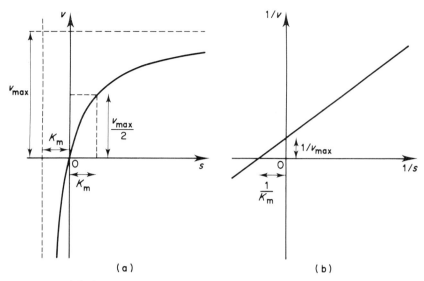

Fig. 4.10 (**a**) Curve of the Michaelis–Menten function $v = v_{max} s/(K_m + s)$. (**b**) The corresponding Lineweaver–Burk line $1/v = 1/v_{max} + (K_m/v_{max})(1/s)$.

temperature, pH, and ionic strength of the buffer solution in which the reaction is occurring, K_m is approximately the dissociation constant of the enzyme–substrate complex. Looked at another way, the reciprocal of K_m, $1/K_m$, is approximately the affinity of an enzyme for its substrate. A high value of K_m, say 0.1 moles per litre, indicates a low enzyme-substrate affinity; and conversely a low value, say 0.0001 moles per litre, shows a high affinity. The vertical asymptote is at $s = -K_m$, but this fact has no biochemical relevance (see Chapter 5, page 65, for the method of calculating asymptotic values).

The Lineweaver–Burk transformation

Just as in the case of the simple rectangular hyperbola, so also may the Michaelis–Menten function be linearly transformed. Inverting both sides of *4.10* gives

$$\frac{1}{v} = \frac{K_m + s}{v_{max}\, s} = \frac{K_m}{v_{max}\, s} + \frac{s}{v_{max}\, s}$$

which, on changing round the terms on the right-hand side, is equivalent to

$$\frac{1}{v} = \frac{1}{v_{max}} + \frac{K_m}{v_{max}} \cdot \frac{1}{s} \qquad (4.11)$$

Relationship 4.11, which is known as the Lineweaver–Burk function, shows that if $1/v$ is plotted against $1/s$ a straight line results of gradient K_m/v_{max} and intercept $1/v_{max}$. Further, put $1/v = 0$ in *4.11*.

$$\text{Then} \qquad \frac{1}{v_{max}} + \frac{K_m}{v_{max}} \cdot \frac{1}{s} = 0$$

$$\text{or} \qquad \frac{K_m}{v_{max}} \cdot \frac{1}{s} = -\frac{1}{v_{max}}$$

The $1/v_{max}$'s cancel, giving $K_m/s = -1$ and so $1/s = -1/K_m$, the negative sign indicating that the intercept on the $(1/s)$-axis is negative (Fig. 4.10b).

As previously mentioned (page 40), it is much easier to work with straight lines than curves; but the Lineweaver–Burk transformation has an added bonus in enzyme studies.

Enzyme inhibition – competitive, non-competitive, and uncompetitive

In a number of instances, adding yet another substance to a substrate solution undergoing an enzyme-controlled reaction results in a diminution of the reaction velocity. The added substance is having an inhibitory effect on the enzymic process.

Enzyme inhibitors are of three kinds. Competitive inhibitors are those which compete with the substance for the active sites on the enzyme, and the effect of the inhibitor can be diminished by increasing substrate concentration. Non-competitive inhibitors may attach to different sites on the enzyme from those occupied by the substrate and/or denature the enzyme. Either of these happenings results in a reduction of activity. A consideration of uncompetitive inhibitors is outside the scope of this account.

Analysis of experimental data by Lineweaver–Burk lines can assist in elucidating the type of inhibition under study.

Experimental results yielding straight lines as in Fig. 4.11(a) indicate competitive inhibition. The higher line in the positive range of $1/s$ is the one in the presence of the inhibitor, because for any given substrate concentration, s, $1/v$ is higher than for the line corresponding to the absence of the inhibitor, and so the velocity, v, is lower. However, both lines have the same intercept, $1/v_{max}$, which implies that both Michaelis–Menten curves have the same maximum velocity at high substrate concentrations. In *equation 4.11*, it is seen that the gradient of a Lineweaver–Burk line is K_m/v_{max}; this implies that in a competitive inhibitory situation only K_m is changed, changed upwards, in fact, showing that under the influence of a competitor, the affinity of the enzyme for its substrate is decreased.

In the case of Fig. 4.11(b), clearly the intercept, $1/v_{max}$, is higher in the presence of the inhibitor than in its absence, showing that the maximum velocity at high substrate concentration, v_{max}, is lower. However, the lines have the same intercept on the $(1/s)$-axis, which shows that they have identical

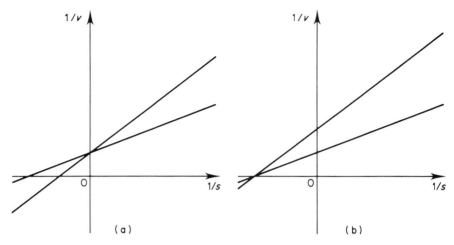

Fig. 4.11 Lineweaver–Burk lines showing: (**a**) competitive inhibition, and (**b**) non-competitive inhibition.

K_m-values. This situation demonstrates that a non-competitive inhibitor does not alter the basic affinity between substrate and enzyme, but that the inhibitor does lower the maximum velocity attainable.

The two situations are shown for the untransformed Michaelis–Menten functions in Fig. 4.12. Clearly, the situation is far easier to assess by the Lineweaver–Burk forms in Fig. 4.11.

The relationship between photosynthetic rate and irradiance

The relationship between the gross photosynthetic rate (p) of a leaf, which can be represented on the vertical axis of a graph, and irradiance (light intensity), I, represented on the horizontal axis, has been found experimentally to be of the form shown by the curve in Fig. 4.13. To quantify the response of photosynthetic rate to irradiance, it is necessary to find a mathematical function which produces such a curve, and one that has been used extensively is

$$p = \frac{1}{a + b/I} \qquad (4.12)$$

There are two constants, a and b, and these will take particular values depending on species and the levels of the other external factors. This function is again a rectangular hyperbola; indeed, it is another version of the Michaelis–Menten function with the two constants in a different form. Thus, the present function may be regarded as a reparameterized version of the

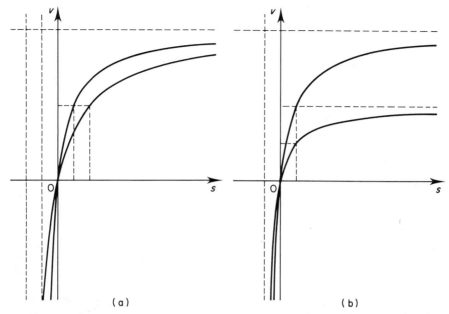

Fig. 4.12 Michaelis–Menten curves corresponding to the Lineweaver–Burk lines in Fig. 4.11. (**a**) Competitive inhibition, and (**b**) non-competitive inhibition.

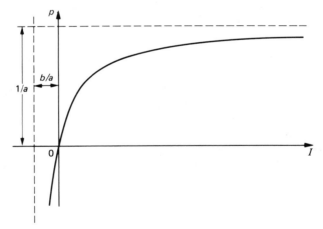

Fig. 4.13 Curve of the function $p = (a + b/I)^{-1}$ relating the photosynthetic rate of a leaf, p, to light intensity, I.

Michaelis–Menten function. It can be shown that the curve also has two asymptotes at right angles to one another: a horizontal one at $p = 1/a$, and a vertical asymptote at $I = -b/a$. The latter is of no biological interest, since it occurs at a negative irradiance, but the asymptote $p = 1/a$ represents the maximum attainable photosynthetic rate as far as irradiance is concerned (commonly known as the light-saturated photosynthetic rate), under the combination of other factors employed in the experiment.

Equation 4.12 shows that, if light is not limiting, then the rate of photosynthesis is given by $1/a$. Evidently, the lower the value of a, the higher will be the maximum rate of photosynthesis, and *vice versa*. Can we give a physical meaning to a?

One way of interpreting the rate of photosynthesis is to consider the rate of movement of carbon dioxide molecules from the air external to the leaf to the place where they are incorporated into the photosynthetic reduction cycle within the chloroplast. A molecule of CO_2 can be imagined to move along a 'path', from air to carboxylation site, and encounter a number of resistances to its movement along the way. First, the molecule moves through the layer of still air adjacent to the leaf surface to a point immediately over the stomatal cavity. This layer of air is known as the boundary layer, and the resistance to molecular movement through the boundary layer is given the symbol r_a. Then the CO_2 molecule passes through the stoma, encountering another resistance, r_s, the stomatal resistance. The molecule now passes from the sub-stomatal cavity, through the air spaces between the leaf mesophyll cells, then through the membranes of a mesophyll cell and the chloroplast to the site of incorporation into the CO_2 reduction cycle. The resistances encountered here are amalgamated to form the so-called mesophyll resistance, r_m. These three resistances are diffusion resistances, but the fourth, called the carboxylation resistance, r_c, is a biochemical one, and controls the rate of entry of the CO_2 molecule into the reduction cycle. Hence there are four resistances in series, and so the total resistance to the passage of the CO_2 molecule is the sum of these. Now since the constant a in *equation 4.12* is inversely related to the maximum photosynthetic rate, it can be identified with the total resistance to the movement of a carbon dioxide molecule from external air to carboxylation site. Thus

$$a = r_a + r_s + r_m + r_c$$

By a different argument, it can be shown that the constant b in *equation 4.12* can also have a physical meaning: it represents the quantum efficiency at low irradiance. Hence the utility of data obtained from experiments which involve the measurement of photosynthetic rate, at differing irradiance, is increased by fitting *equation 4.12* to such data, and then making comparisons of the constants obtained between species or within the same species under different regimes of other environmental factors.

EXERCISES

1. Evaluate (a) $f(0)$ (b) $f(2)$ (c) $f(-2)$

$$\text{for} \quad f(x) = 2x^3 - 3x^2 + 4x + 1$$

2. Evaluate (a) $f(4, 1)$ (b) $f(-1, -2)$

$$\text{for} \qquad f(x, y) = 2x^2 + 3xy - y^2 - 3x - 2y + 5$$

3. Draw a graph of the function $p = (3 + 0.5/I)^{-1}$ in the range $0 \leqslant I \leqslant 1.2$. The function represents the relationship between gross photosynthetic rate, p (mg CO_2 absorbed $cm^{-2}\, h^{-1}$), and irradiance, I (cal $cm^{-2}\, min^{-1}$) in the leaf of a certain plant species. What is the photosynthetic rate when I is:

(a) 0.1   and   (b) 1.0 cal $cm^{-2}\, min^{-1}$?

4. The enzyme maltase, which catalyses the conversion of p-nitrophenyl-α-D-glucoside to D-glucose and p-nitrophenol, is inhibited by glucose. In an experiment, the following values of the parameters of Michaelis–Menten functions were estimated from the data: in the absence of glucose, $v_{max} = 0.2174$ μmol min^{-1}, $K_m = 0.25$ mM; in the presence of glucose, $v_{max} = 0.2174$ μmol min^{-1}, $K_m = 0.345$ mM. Draw a graph of each Lineweaver–Burk function on the same axes in the range $-5 < 1/s < 35$ $(mM)^{-1}$. Is the inhibitory action of glucose competitive or non-competitive in this system? Calculate the velocity of the reaction, in each case, when substrate concentration is 0.1 mM.

5

Differentiation

This chapter and the two and a half succeeding ones will introduce the branch of mathematics known as the calculus. School mathematics is often rigidly divided into arithmetic, algebra, geometry, and trigonometry, but, as the study of mathematics is taken to deeper levels, these somewhat artificial divisions are steadily eroded. These sub-divisions are fairly natural for basic mathematics, but the calculus does not rank with them. The calculus is essentially a tool – a set of mathematical methods invented by Sir Isaac Newton. Problems involving the use of calculus *could* be solved by other methods, but with much greater difficulty and labour.

Calculus deals basically with rates of change, and the facility with which it does this means that it is eminently suited to dealing with dynamic situations as opposed to static ones. All living material is continually in a state of change, and so it follows that the calculus is an indispensable tool in biomathematics: it aids in formulating and solving many problems, such as those concerning growth, biochemical changes, and population dynamics.

The method of the calculus is to discover properties of complete entities by examining the behaviour of their elementary components. An example of the latter, in physics, is in studying the mode of vibration of a stretched string, which we do by considering the motion of each particle of the string independently and then combining all these motions together. This may sound an extremely tedious way to study the vibration of a string, as indeed it would be without the invention of the calculus. By calculus methods, however, the problem can be formulated and solved in a few lines.

Calculus as a subject contains two subdivisions: the differential calculus and the integral calculus. The former is concerned with calculating the rate of change of a function with respect to its variable, e.g. the rate of change of $f(x)$ with respect to x. The latter subdivision of calculus is concerned with the reverse process: given the rate of change of a function with respect to its variable, what is the function itself? The integral calculus also provides techniques for adding together the results obtained for individual components of an entity when we are considering the behaviour of the complete entity.

In this chapter the principles of differentiation, and methods for differentiating mathematical functions are described; but first, a more general mathematical concept must be introduced.

Limits

Limits involving infinity

To approach the idea of a mathematical limit, consider again the subject of asymptotes, introduced in the previous chapter. There, we examined the behaviour of the function $y = 1/x$ through a study of its curve, and found that as x increased indefinitely, y approached nearer and nearer to zero. But however large x might be, y was never actually equal to 0.

Let us for a moment imagine that there could be an extremely large number, called infinity, which is given the symbol ∞. It is then intuitively evident that $y = 1/\infty = 0$. That there cannot be an extremely large number called ∞ is obvious from the fact that however large a number one can specify, a still larger number can always be named. Numbers continue increasing without end.

What we can say, however, is that as x increases, $1/x$ decreases, and if x *could* reach infinity $1/x$ *would* be zero. In less imaginary terms, we say that the *limit* of $1/x$, as x approaches infinity, is zero; and the statement is written in symbols as

$$\lim_{x \to \infty} \left(\frac{1}{x} \right) = 0 \qquad (5.1)$$

Another way of viewing the limit can also be seen in this example. As $x \to \infty$, $1/x$ is always positive but decreasing in magnitude; $1/x \to 0$, but never goes below 0, i.e. never starts acquiring negative values. We have already seen this (Fig. 4.8a) when considering the curve of $y = 1/x$; this curve approaches the x-axis (i.e. the line $y = 0$) as x increases, but never intersects it. Such a line is an asymptote, and using the definition of an asymptote (page 54) we infer that the line $y = 0$ will 'touch the curve at infinity'.

Example 5.1

Find the asymptotes to the curve of the function describing the response of photosynthetic rate, y, to light intensity, x

$$y = \frac{1}{a + (b/x)} \qquad (5.2)$$

This question implies finding the limit of y as x approaches ∞, and the limit of x as y approaches ∞. In the first case let $x \to \infty$ in the above equation. Evidently $(b/x) \to 0$, and so $y \to 1/a$.

$$\text{Hence} \qquad \lim_{x \to \infty} \left(\frac{1}{a + (b/x)} \right) = \frac{1}{a} \qquad (5.3)$$

To find the other limit, we first obtain x in terms of y. From *5.2* by cross-multiplication:

$$a + \frac{b}{x} = \frac{1}{y}$$

or
$$\frac{b}{x} = \frac{1}{y} - a$$

and so
$$x = \frac{b}{(1/y) - a}$$

As $y \to \infty$, $1/y \to 0$ and so $x \to -b/a$. Hence we may write

$$\lim_{y \to \infty} \left(\frac{b}{(1/y) - a} \right) = -\frac{b}{a} \qquad (5.4)$$

Thus the asymptotes to the curve of *equation 5.2* are the horizontal line $y = 1/a$ and the vertical line $x = -b/a$ (see Fig. 4.13).

Limits not involving infinity

Limits of functions can be obtained when the variable approaches values other than infinity, as the following examples show.

Example 5.2
   Find the limit of $(3 - x)/(2 + x)$ as

$$(a)\ x \to 2, \quad (b)\ x \to 3, \quad (c)\ x \to -2$$

(a) Examining the effect of letting x approach 2 in the numerator and denominator separately, we see that as $x \to 2$, $(3 - x) \to 1$ and $(2 + x) \to 4$.

$$\text{Thus} \qquad \lim_{x \to 2} \left(\frac{3 - x}{2 + x} \right) = \frac{1}{4}$$

(b) Adopting the same approach as in (a), we see that as $x \to 3$, $(3 - x) \to 0$ and $(2 + x) \to 5$.

$$\text{Hence} \qquad \lim_{x \to 3} \left(\frac{3 - x}{2 + x} \right) = 0$$

(c) Here, as $x \to -2$, $(3 - x) \to 5$ and $(2 + x) \to 0$;

$$\text{and so} \qquad \lim_{x \to -2} \left(\frac{3 - x}{2 + x} \right) = \infty$$

What this result means in words is that the limit is an indefinitely large number.

In *example 5.2(a)* and *(b)*, the limits are easy to evaluate. This is because, rather than saying what happens to the expression $(3 - x)/(2 + x)$ when $x \rightarrow 2$ or $x \rightarrow 3$, we can see the result of putting $x = 2$ and $x = 3$. The value of $(3 - x)/(2 + x)$ when $x = 2$ is $\frac{1}{4}$, and when $x = 3$ the value of the function is $\frac{0}{5} = 0$. Things are not quite so straightforward in the case of *example 5.2(c)* since, by putting $x = -2$, $(3 - x)/(2 + x) = \frac{5}{0}$ which is indeterminate. However, remembering that a quotient with a very small number in the denominator yields a large number, then we can say that as the denominator *approaches* zero, the quotient itself must get very large. So, when the denominator becomes *vanishingly* small, the whole fraction becomes infinitely large.

Example 5.3
Find the limits of

$$(a) \quad \frac{x^2 - 2x}{x} \qquad \text{as} \qquad x \rightarrow 0$$

$$(b) \quad \frac{x^2 - 1}{x - 1} \qquad \text{as} \qquad x \rightarrow 1$$

(a) If x is put equal to 0, then $(x^2 - 2x)/x = \frac{0}{0}$, which is completely indeterminate. Since both numerator and denominator are zero in these circumstances, it is not possible to proceed as in *example 5.2(c)*.

One way of proceeding is as follows. The square of a very small number is smaller than the number itself. So if x is very small, x^2 is negligible compared with x, and so the fraction reduces to $-2x/x = -2$. The smaller x is, the more accurate does the approximation (i.e. ignoring x^2) become; so that when x is vanishingly small, we may write

$$\lim_{x \to 0} \left(\frac{x^2 - 2x}{x} \right) = -2$$

A much better method is to divide out the fraction first; then we have, directly

$$\lim_{x \to 0} \left(\frac{x^2 - 2x}{x} \right) = \lim_{x \to 0} (x - 2) = -2$$

(b) We note that $x^2 - 1 = (x - 1)(x + 1)$.

$$\text{So} \quad \lim_{x \to 1} \left(\frac{x^2 - 1}{x - 1} \right) = \lim_{x \to 1} \left\{ \frac{(x - 1)(x + 1)}{x - 1} \right\} = \lim_{x \to 1} (x + 1) = 2$$

Examples 5.3(a) and *(b)* are particularly relevant to the underlying principle of the differential calculus. It will be observed in each of these two examples that *although the numerator and denominator of the fractions separately tend to zero as x tends to the value shown, the entire quotient remains finite and tends to a finite value.*

It must be emphasized that this very brief introduction to limits has been to aid in understanding the principle of differentiation, rather than as an end in itself. The subject of limits in mathematics is a large and often difficult one, and we have approached it in a rather empirical fashion.

The principle of differentiation

Consider the graph of any function, $y = f(x)$ (Fig. 5.1), and let $P(x, y)$ be any point on the curve. Let a point Q be situated close to P (the distance is highly magnified in Fig. 5.1 for clarity) such that the horizontal distance between P and Q is the interval δx, and the vertical distance between the two points is the interval δy. Thus the co-ordinates of point Q are $(x + \delta x, y + \delta y)$.

In mathematical notation, an interval (that is, a distance between points), is usually denoted by the capital Greek letter delta, Δ, but if the interval is small the equivalent lower case (small) letter is employed, δ. Thus an interval on an x-scale is Δx, but if the interval is small it is denoted by δx. Note that neither Δ nor δ are algebraic quantities, but are a kind of prefix. In the term Δx, Δ and x can never be separated.

With reference again to Fig. 5.1, the straight line PQ is known as a **chord** of the curve, and its gradient is obviously $\delta y/\delta x$. The point made in the previous paragraph, that δ is not an algebraic quantity but a prefix, applies here. There is no question of the δ's cancelling out; the term $\delta y/\delta x$ means a small interval in y divided by a small interval in x.

Now let point Q move along the curve towards point P. When this happens δx (and δy), already small, decrease in size – approach zero in fact; but although δx and δy both approach zero, their quotient remains finite (see above). Evidently the gradient of PQ will change under these circumstances and

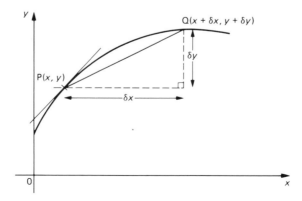

Fig. 5.1   The principle of differentiation.

the chord PQ will, in fact, approach the tangent at P. Hence, when δx and δy become vanishingly small, *but not zero*, the chord PQ will be indistinguishable from the tangent at point P; more important, the *gradient* of chord PQ will be indistinguishable from the *gradient* of the tangent at P under these conditions.

The gradient of the tangent at P is the limit of the gradient of the chord PQ as PQ becomes vanishingly small, and is denoted by dy/dx;

$$\text{so} \quad \lim_{\delta x \to 0} \left(\frac{\delta y}{\delta x} \right) = \frac{dy}{dx} \quad (5.5)$$

In 5.5, $\delta y/\delta x$ is the gradient of the chord and dy/dx is the gradient of the tangent as already defined; so 5.5 can be expressed in words as, 'the limit of the gradient of a chord, $\delta y/\delta x$, as δx tends to zero, is the gradient of a tangent, dy/dx'. In the term dy/dx, the d's are again prefixes, denoting vanishingly small intervals.

Differentiation from first principles

To fix our ideas, let us differentiate some of the functions we encountered in the previous chapter.

Example 5.4
Differentiate the following functions with respect to x.

$$(a)\, y = x^3, \quad (b)\, y = 1/x$$

(a) This is a very simple cubic polynomial containing only a term in x^3 with a coefficient of 1. The curve is shown in Fig. 5.2, and P is any point on the curve with co-ordinates (x, y). Clearly there is a relationship between x and y, the functional relationship

$$y = x^3 \quad (5.6)$$

Consider a point Q, which is reached by moving a small distance along the curve from P so that its position is δx from P in the direction of the x-axis, and a corresponding small increment (δy) of y from P. Point Q will have co-ordinates $(x + \delta x, y + \delta y)$, and again there is a relationship between the two co-ordinates:

$$y + \delta y = (x + \delta x)^3 \quad (5.7)$$

Now subtract 5.6 from 5.7

$$(y + \delta y) - y = (x + \delta x)^3 - x^3$$

$$\text{i.e.} \quad \delta y = (x + \delta x)^3 - x^3$$

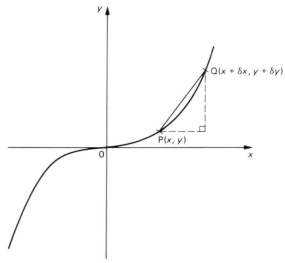

Fig. 5.2   Application of the principle of differentiation to the function $y = x^3$ (see *example 5.3a*).

Next, expand the cubed bracket, so that we have

$$\delta y = x^3 + 3x^2\,\delta x + 3x(\delta x)^2 + (\delta x)^3 - x^3$$

in which the two x^3 terms cancel one another out, and we are left with

$$\delta y = 3x^2\,\delta x + 3x(\delta x)^2 + (\delta x)^3$$

On dividing both sides of this expression by δx, we have

$$\frac{\delta y}{\delta x} = \frac{3x^2\,\delta x}{\delta x} + \frac{3x(\delta x)^2}{\delta x} + \frac{(\delta x)^3}{\delta x}$$

and, on cancelling δx's on the right-hand side:

$$\frac{\delta y}{\delta x} = 3x^2 + 3x\delta x + (\delta x)^2 \tag{5.8}$$

Equation 5.8 is an expression by means of which the gradient of chord PQ can be evaluated for a given x and δx.

Now let δx tend to zero. This means that δx becomes so small that it can be taken as zero in *expression 5.8*, so that the second and third terms on the right-hand side of that equation disappear, leaving only $3x^2$. The quotient $\delta y/\delta x$ on the left-hand side thus remains finite and tends to a finite limit, $3x^2$, which we again denote as dy/dx (as in *equation 5.5*). So, in the present example

$$\lim_{\delta x \to 0} \{3x^2 + 3x\delta x + (\delta x)^2\} = 3x^2 \qquad (5.9)$$

On comparing *5.5* and *5.9*, we see that $dy/dx = 3x^2$. Hence, for the function $y = x^3$, $dy/dx = 3x^2$. This means that the gradient of the tangent to the curve $y = x^3$, at a point whose x-co-ordinate is x, is given by $3x^2$.

(*b*) This is the simplest rectangular hyperbola (page 55), and one part of the curve is reproduced in Fig. 5.3. The differentiation procedure is exactly the same as in the previous example, so only the outline will be given here. At point P, we have the relationship

$$y = \frac{1}{x} \qquad (5.10)$$

At point Q, we have the similar relationship

$$(y + \delta y) = \frac{1}{(x + \delta x)} \qquad (5.11)$$

where δy in this instance is a negative quantity (see Fig. 5.3).
5.11 minus *5.10* gives

$$\delta y = \frac{1}{(x + \delta x)} - \frac{1}{x}$$

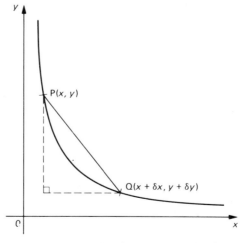

Fig. 5.3 Application of the principle of differentiation to the function $y = 1/x$; δy is a negative quantity (see *example 5.4b*).

Put the right-hand side over a common denominator:

$$\delta y = \frac{x - (x + \delta x)}{x(x + \delta x)} = \frac{-\delta x}{x(x + \delta x)}$$

Dividing both sides by δx gives

$$\frac{\delta y}{\delta x} = \frac{-1}{x(x + \delta x)} \tag{5.12}$$

As before, *equation 5.12* gives the gradient of the chord PQ for a given x and δx. Hence it is evident that

$$\lim_{\delta x \to 0} \left\{ -\frac{1}{x(x + \delta x)} \right\} = -\frac{1}{x^2} \tag{5.13}$$

Again, comparing *5.5* and *5.13*, we see that $dy/dx = -1/x^2$.

Differentiation of polynomial functions

The rule for differentiating $y = ax^n$

In the first example above, we found that if $y = x^3$, then $dy/dx = 3x^2$, i.e. $3x^{(3-1)}$. In the second example, $y = 1/x$ and $dy/dx = -1/x^2$. Now $1/x$ can be written as x^{-1}, and $-1/x^2$ as $-x^{-2}$; thus we have that when $y = x^{-1}$, $dy/dx = -x^{-2}$, i.e. $-1x^{(-1-1)}$. We see that in both of these examples we are differentiating an expression of the form $y = x^n$, where $n = 3$ in the first case, and $n = -1$ in the second. In both instances, the result for dy/dx is obtained if x is multiplied by n, and 1 is deducted from n in the exponent. Thus, if

$$\left. \begin{array}{c} y = x^n \\[2ex] \text{then} \quad \dfrac{dy}{dx} = nx^{(n-1)} \end{array} \right\} \tag{5.14}$$

The two expressions in *5.14* in fact state a general rule, which applies if n is any positive or negative integer.

Now suppose that x^n is multiplied by a constant a (i.e. by a coefficient), giving ax^n. It can easily be demonstrated (Exercise 1 at the end of this chapter) that if $y = ax^n$, $dy/dx = anx^{(n-1)}$. In other words, the coefficient, a, remains unchanged, and does not affect the operation of the basic rule.

Example 5.5
Differentiate the following functions with respect to x, by the rule provided above.

$$(a)\, y = 5x^2 \quad (b)\, y = -3x^5 \quad (c)\, y = \frac{2}{x^2}$$

(*a*) We can immediately write $\dfrac{dy}{dx} = (5)2x^{(2-1)} = 10x$

(*b*) Again, we have $dy/dx = (-3)5x^{(5-1)} = -15x^4$

(*c*) Write the expression as $y = 2x^2$.

$$\text{Then} \qquad \frac{dy}{dx} = (2)(-2)x^{(-2-1)} = -4x^{-3}$$

$$\text{i.e.} \qquad \frac{dy}{dx} = -\frac{4}{x^3}$$

Differentiation of roots

Although it will not be proved here, the rule in *equation 5.14* also applies when the exponent, n, is a fraction. If we recall that a fractional exponent implies a root (page 11), e.g. $x^{1/2} = \sqrt{x}$, this means that we can use the rule to differentiate roots of x as well as x raised to powers.

Example 5.6

Differentiate the following expressions with respect to x.

$$(a)\; y = \sqrt[4]{x} \quad (b)\; y = 3\sqrt{x} \quad (c)\; y = 1/\sqrt[3]{x} \quad (d)\; y = \sqrt{(x^3)}$$

(*a*) Write the expression as $y = x^{1/4}$

$$\text{then} \qquad \frac{dy}{dx} = \tfrac{1}{4}x^{(1/4-1)}$$

$$\text{i.e.} \qquad \frac{dy}{dx} = \tfrac{1}{4}x^{-3/4} = \frac{1}{4x^{3/4}}$$

$$\text{Thus} \qquad \frac{dy}{dx} = \frac{1}{4\sqrt[4]{(x^3)}}$$

(*b*) $y = 3x^{1/2}$, then $\dfrac{dy}{dx} = 3(\tfrac{1}{2})x^{(1/2-1)} = \tfrac{3}{2}x^{-1/2}$

$$\text{i.e.} \qquad \frac{dy}{dx} = -\frac{3}{2x^{1/2}} = -\frac{3}{2\sqrt{x}}$$

(c) $y = 1/x^{1/3}$,   i.e. $y = x^{-1/3}$

            Hence $dy/dx = (-\tfrac{1}{3})x^{-4/3} = -1/(3x^{4/3})$

            Thus $dy/dx = -1/\{3\sqrt[3]{(x^4)}\}$

(d) $y = x^{3/2}$,   hence $dy/dx = \tfrac{3}{2}x^{1/2} = \tfrac{3}{2}\sqrt{x}$

Differentiation of a constant

The result is most easily demonstrated by means of an example.

Example 5.7
Differentiate $y = a$ (where a is constant), with respect to x. It is implied that y is a function of x even though x does not actually appear in the function; however, the expression could be written as $y = ax^0$, which now clearly

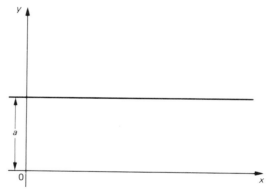

Fig. 5.4   The straight line $y = a$.

demonstrates that y can be regarded as a function of x. This means that we can draw a graph of the function $y = a$ on an x–y plane, and this is shown in Fig. 5.4. The 'curve' of $y = a$ is a horizontal straight line, and we can differentiate from first principles in the usual way by considering two points on the line, $P(x,y)$ and $Q(x + \delta x, y + \delta y)$. For P, $y = a$; and for Q, $y + \delta y = a$. Subtracting, we have $\delta y = 0$, and so $\delta y/\delta x = dy/dx = 0$. So, ***differentiating a constant yields zero.*** This is to be expected, since the gradient of a horizontal line is zero.

Differentiation of a polynominal function

A polynomial function of x consists of a series of terms of the form ax^n added together. Can we differentiate a polynomial by dealing with each term separately? *Example 5.8* shows that we can.

Example 5.8

Differentiate $y = x^2 - 5x + 2$ with respect to x from first principles.

In the usual way, formulate two equations for points P and Q which lie on the curve of the function.

$$y = x^2 - 5x + 2 \qquad (5.15)$$

$$y + \delta y = (x + \delta x)^2 - 5(x + \delta x) + 2 \qquad (5.16)$$

Subtraction of *5.15* from *5.16* gives

$$\delta y = (x + \delta x)^2 - 5(x + \delta x) - x^2 + 5x$$

Expand the brackets

$$\delta y = x^2 + 2x\delta x + (\delta x)^2 - 5x - 5\delta x - x^2 + 5x$$

i.e. $\qquad \delta y = 2x\delta x + (\delta x)^2 - 5\delta x$

Divide both sides by δx

$$\frac{\delta y}{\delta x} = 2x + \delta x - 5$$

and so $\qquad \lim_{\delta x \to 0} (2x + \delta x - 5) = 2x - 5 = \dfrac{dy}{dx}$

If *5.15* is differentiated term by term using the rule in *5.14*, we obtain the same result as in *example 5.8*. Hence, a polynomial function can be differentiated by applying the rule for differentiating ax^n to each term in the polynomial separately.

The differential coefficient (derivative)

We have developed the discussion of differentiation by depicting the various mathematical activities on graphs; thus, we have given dy/dx a meaning: it is the gradient of the tangent to a curve, and hence the gradient of the curve itself at that point (see page 69). The differentiation of a function can, however, be accomplished without any reference to a curve, and without giving the result a physical meaning. It is not even necessary to invoke a second variable, e.g. y; we can work entirely in terms of the one variable x. For example, we have already seen that if $y = x^n$, then $dy/dx = nx^{(n-1)}$; but we could embody both these expressions into one, and write

$$\frac{d(x^n)}{dx} = nx^{(n-1)} \qquad (5.17)$$

Instead of y in the numerator of the left-hand side, we have written what y is in terms of x, namely x^n. So *example 5.8* could have been stated, 'Differentiate $x^2 - 5x + 2$ with respect to x from first principles', or more succinctly, 'Find $d(x^2 - 5x + 2)/dx$ from first principles'.

Reverting to two variables for a moment, consider the case where y is an unknown function of x, and we write $y = f(x)$. Then the differentiated function (which is also unknown) is written as $dy/dx = f'(x)$. This notation can also be used when the second variable, y, is not employed; e.g. if $f(x) = x^2 - 5x + 2$, then $f'(x) = 2x - 5$.

If a function of x, $f(x)$, is differentiated, we obtain a new function of x which we write as $f'(x)$. The function $f'(x)$ is *derived* from the original function, $f(x)$, by the process of differentiation, and $f'(x)$ may be called the **derivative** of $f(x)$ but is also known as the **differential coefficient**. Thus, for example, $3x^2$ is the differential coefficient (or derivative) of x^3.

Differentiation of more complicated functions

Differentiation of a sum or difference of terms

So far, we have differentiated from first principles most of those functions that were introduced in the previous chapter, namely, the polynomials, and the simple rectangular hyperbola.

In the case of polynomials, it was found that each term could be differentiated separately. As an example for further discussion, consider again the polynomial function $y = x^2 - 5x + 2$. Now the term 'function' can be used at many different levels of complexity. For instance, in this example if we let $y = f(x)$, then $f(x) = x^2 - 5x + 2$. But each of the terms on the right-hand side are also functions of x; even the last term is a function of x as it could be written as $2x^0$. Thus, we could put $F(x) = x^2$, $\phi(x) = -5x$, and $\psi(x) = 2$; and so

$$f(x) = F(x) + \phi(x) + \psi(x) \qquad (5.18)$$

In other words, a function of x that is a polynomial is made up of a number of simple functions of x added together.

We have already seen that $f'(x) = 2x - 5$ for this example (page 75). Differentiating the individual functions on the right-hand side of *5.18*, we find $F'(x) = 2x$, $\phi'(x) = -5$, and $\psi'(x) = 0$. So again we have the result that $f'(x) = 2x - 5$. This means that we can formulate the following rule.

The derivative of a function which consists of a number of terms added together is equal to the sum of the derivatives of each term.

The term 'addition' also includes 'subtraction'. In symbols, for just two terms, if

$$\left. \begin{array}{l} f(x) = \phi(x) \pm \psi(x) \\ \text{then} \quad f'(x) = \phi'(x) \pm \psi'(x) \end{array} \right\} \qquad (5.19)$$

Differentiation of a product

Suppose we have the following function of x to differentiate:
$$f(x) = (x^3 - 2)(x + 1)$$
One way of doing it would be to multiply out the brackets first, which would give a polynomial, and then differentiate the latter in the usual way. This example as originally given, however, is obviously made up of two functions of x multiplied together. If we let $\phi(x) = x^3 - 2$ and $\psi(x) = x + 1$, then
$$f(x) = \phi(x)\psi(x)$$
Now, just as we found above that there is a rule for differentiating a sum or a difference of two functions of x, so there is a rule to differentiate a product of two functions of x, and it is often much more convenient to apply than it is to multiply out the given product first. The rule, which we shall not prove here, is rather cumbersome in words; so in symbols, if

$$\left. \begin{array}{ll} & f(x) = \phi(x)\psi(x) \\ \text{then} & f'(x) = \phi(x)\psi'(x) + \psi(x)\phi'(x) \end{array} \right\} \tag{5.20}$$

Example 5.9
Find $\quad d\{(x^3 - 2)(x + 1)\}/dx$

In the notation of *5.20*, we have

$$\phi(x) = x^3 - 2 \quad \text{and} \quad \psi(x) = x + 1$$

$$\text{so} \quad \phi'(x) = 3x^2 \quad \text{and} \quad \psi'(x) = 1$$

$$\text{Hence} \quad f'(x) = (x^3 - 2)(1) + (x + 1)(3x^2)$$

$$\text{i.e.} \quad f'(x) = 4x^3 + 3x^2 - 2$$

We can check this result by multiplying out the brackets first:

$$f(x) = x^4 + x^3 - 2x - 2 \quad \text{and so} \quad f'(x) = 4x^3 + 3x^2 - 2$$

Differentiation of a quotient

In principle, the rule for the differentiation of a quotient is similar to that for a product. If

$$\left. \begin{array}{l} f(x) = \phi(x)/\psi(x) \\[2mm] f'(x) = \dfrac{\psi(x)\phi'(x) - \phi(x)\psi'(x)}{\{\psi(x)\}^2} \end{array} \right\} \tag{5.21}$$

Example 5.10
  Find
$$d\{(x^3 - 2)/(x + 1)\}/dx$$

$$\phi(x) = x^3 - 2 \quad \text{and} \quad \psi(x) = x + 1$$

Hence $\quad \phi'(x) = 3x^2 \quad \text{and} \quad \psi'(x) = 1$

then $\quad f'(x) = \dfrac{3x^2(x + 1) - (x^3 - 2)}{(x + 1)^2} = \dfrac{2x^3 + 3x^2 + 2}{(x + 1)^2}$

Function of a function

Suppose we have to differentiate with respect to x the function $(5x^2 + 7x + 2)^6$. There are three ways in which this could be done: (*i*) expand the bracket and then differentiate the resulting polynomial term by term (excessively laborious, and very liable to error); (*ii*) write out the bracketed term six times and then use an extension of the product rule (still very laborious); or (*iii*) use a new rule called the function of a function (very much quicker), as described in the following example.

Example 5.11
  Find
$$d\{(5x^2 + 7x + 2)^6\}/dx$$

When using the function of a function rule, it is convenient to start by putting the function to be differentiated equal to another variable, if this is not already specified;

$$\text{so let} \quad y = (5x^2 + 7x + 2)^6 \tag{5.22}$$

The procedure then is to put some suitable part of the function, which can be differentiated by some other method, equal to yet another variable.

$$\text{In this example let} \quad u = 5x^2 + 7x + 2 \tag{5.23}$$

then *5.22* can be written as

$$y = u^6 \tag{5.24}$$

Now differentiate *5.24* with respect to u:

$$\frac{dy}{du} = 6u^5 \tag{5.25}$$

and differentiate *5.23* with respect to x:

$$\frac{du}{dx} = 10x + 7 \tag{5.26}$$

The function of a function rule states that, in the present notation,

$$\frac{dy}{dx} = \frac{dy}{du} \cdot \frac{du}{dx} \qquad (5.27)$$

We see from *5.22* that the answer to the original problem will be given by dy/dx — the left-hand side of *5.27*; and it is evident that we already have the two terms on the right-hand side of *5.27*, in *5.25* and *5.26*. Hence

$$\frac{dy}{dx} = (6u^5)(10x + 7) \qquad (5.28)$$

Expression 5.28 is not, however, quite the final answer. The original function contained only the variable x, and the answer must do so too. So, first eliminating u using *5.23*, we have

$$\frac{dy}{dx} = 6(5x^2 + 7x + 2)^5(10x + 7)$$

and so, finally, we have that

$$\frac{d\{(5x^2 + 7x + 2)^6\}}{dx} = 6(5x^2 + 7x + 2)^5(10x + 7)$$

Example 5.12
Differentiate the following functions with respect to x.

$$(a) \ 1/\sqrt{(2x^3 + 4)} \qquad \text{and} \qquad (b) \ x\sqrt{(2x + 1)}$$

(a) At first sight, it might be thought that the quotient rule would be appropriate here. But when differentiation of the denominator is attempted, the function of a function rule has to be used. It is more convenient to use the function of a function rule only, since the given function can be written as $(2x^3 + 4)^{-1/2}$ using the laws of indices.

So, let $\qquad y = (2x^3 + 4)^{-1/2}$

and let $\qquad u = 2x^3 + 4$

giving $\qquad y = u^{-1/2}$

Now $\qquad \dfrac{dy}{du} = -\tfrac{1}{2}u^{-3/2}$

and $\qquad \dfrac{du}{dx} = 6x^2$

Thus $\dfrac{dy}{dx} = (-\tfrac{1}{2}u^{-3/2})(6x^2) = -3x^2u^{-3/2} = -\dfrac{3x^2}{\sqrt{(u^3)}}$

and so $\dfrac{d\{1/\sqrt{(2x^3+4)}\}}{dx} = -\dfrac{3x^2}{\sqrt{\{(2x^3+4)^3\}}}$

(*b*) This example is a product of two functions of x; i.e. x, and $(2x+1)^{1/2}$. The latter, however, needs to be differentiated by the function of a function rule first.

Let $y = (2x+1)^{1/2}$

and let $u = 2x+1$

giving $y = u^{1/2}$

Now $\dfrac{dy}{du} = \tfrac{1}{2}u^{-1/2}$

and $\dfrac{du}{dx} = 2$

Thus $\dfrac{dy}{dx} = (\tfrac{1}{2}u^{-1/2})(2) = u^{-1/2}$

So $\dfrac{d\{\sqrt{(2x+1)}\}}{dx} = \dfrac{1}{\sqrt{(2x+1)}}$

Now apply the product rule to the whole expression:

$\dfrac{d\{x\sqrt{(2x+1)}\}}{dx} = \dfrac{x}{\sqrt{(2x+1)}} + \sqrt{(2x+1)} = \dfrac{3x+1}{\sqrt{(2x+1)}}$

Higher derivatives

Imagine that we have the following quadratic function

$$y = 5x^2 + 7x + 2 \qquad\qquad (5.29)$$

Differentiation gives $\dfrac{dy}{dx} = 10x + 7 \qquad\qquad (5.30)$

Suppose, however, we had been given the function

$$y = 10x + 7 \qquad (5.31)$$

Differentiation of this gives $\qquad \dfrac{dy}{dx} = 10 \qquad\qquad (5.32)$

Now clearly, the actual functions of x involved in the right-hand sides of expressions *5.30* and *5.31* are identical. The only difference is that *5.30* represents the differential coefficient of *5.29*, while *5.31* has not been written here as a differential coefficient. But obviously the function $10x + 7$ can be differentiated with respect to x, and yields 10. Is there any reason why we should not say that differentiation of *5.29* gives *5.30* and that differentiation of *5.30* gives *5.32*? No reason at all: in fact $5x^2 + 7x + 2$ differentiated gives $10x$ + 7, and when differentiated again gives 10. In symbols:

$$y = 5x^2 + 7x + 2 \qquad \text{or} \qquad f(x) = 5x^2 + 7x + 2$$

$$\frac{dy}{dx} = 10x + 7 \qquad \text{or} \qquad f'(x) = 10x + 7$$

$$\frac{d^2y}{dx^2} = 10 \qquad \text{or} \qquad f''(x) = 10$$

The equations on the left above give the notations when the situation warrants the use of two variables, y being a function of x, and the equations on the right show the symbols used when we have no desire to mention another variable, y.

It has been stated that dy/dx $(= f'(x))$ is called the derivative, or differential coefficient. This terminology now requires expansion: we say that dy/dx $(= f'(x))$ is the **first derivative**, or the **first differential coefficient**. Likewise, d^2y/dx^2 (which in words is read as, 'd-two-y by d-x-squared', and not as, 'd-squared-y etc.') and $f''(x)$ denote the **second derivative**, or the **second differential coefficient**.

There is, in fact, nothing to stop us obtaining differential coefficients of higher order than the second, but as they are of no practical value to the biologist, we shall leave the subject at this point.

EXERCISES

1. By differentiating from first principles, show that

$$\frac{d(ax^4)}{dx} = 4ax^3$$

In exercises 2 and 3, differentiate the functions with respect to x.

2. (a) $(3x - 2)^3$ (b) $1/(ax^2 + bx + c)$ (c) $(x + 1/x)^n$

3. (a) $(1 + x^2)/(1 - x^2)$ (b) $x/\sqrt{(1 - x)}$ (c) $x^2\sqrt{(a^2 - x^2)}$
   (d) $\{(ax + b)/(cx + d)\}^n$

4. Find dy/dx and d^2y/dx^2 for the functions

$$(a)\, y = 1/x^2 \quad \text{and} \quad (b)\, y = \frac{ax}{b + x}$$

6

Use of the differential calculus

The physical meanings of the first and second derivatives

Before any mathematical concept can be put to use, the symbols and procedures involved must be identified with quantities and processes which occur in the physical world. So far, we have seen that the first derivative of a function represents the gradient of the curve of that function, but we have not yet attempted to assign any kind of meaning to the second derivative.

Now a mathematical function $f(x)$ is an abstract entity. For any value of x, $f(x)$ can *usually* be evaluated; but this is not always possible: consider, for instance, the function $1/x$ when $x = 0$. Leaving aside these kinds of difficulties, which are rare in practice, we can say that given a numerical value of x, a numerical value of $f(x)$ can be calculated. If this is done for a number of different x-values, we could write the results in a table. Now x and $f(x)$ are obviously variable quantities which we can identify with the axes of a two-dimensional graph, x for the horizontal axis – the independent variable – because it can take any value we please, and $f(x)$ for the vertical axis – the dependent variable – because its values depend on those of x. Often, $f(x)$ is given a single symbol of its own, y, but this is not necessary.

By using the tabulated values of $f(x)$ for each x, a curve may be drawn with respect to the two axes, and this curve is a *physical representation* of the function. Again, the first derivative, $f'(x)$, is a purely abstract concept, derived from the original function, $f(x)$, which is also abstract; but we have already shown that $f'(x)$ represents the gradient of the curve of $f(x)$, which gives the first derivative a physical meaning.

However, the foregoing discussion, although a step in the right direction, is not sufficient for us to be able to use the differential coefficient in order to solve problems arising in the real world. We now need to identify features connected with the curve of a function with happenings in the physical world, and a simple example is to consider the motion of a body, say a marble, in a straight line along the top of a bench.

Distance and velocity – the first derivative

Let us imagine a marble at rest at zero time ($t = 0$) on one end of the bench (zero distance, $s = 0$). Assume now that the marble starts moving. For this to

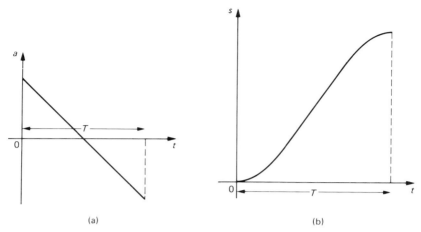

Fig. 6.1   (**a**) An acceleration–time curve whose equation is $a = 3(T - 2t)$. (**b**) A portion of the corresponding distance–time curve whose equation is $s = 3Tt^2/2 - t^3$.

happen, the marble must, of course, accelerate, but it is not necessary here to enquire into the force that must be applied to the marble to cause it to accelerate.

We shall assume that the initial acceleration decreases linearly with time, and the situation is shown in Fig. 6.1(a). The effect can be likened to a car driver very gradually easing his foot off the accelerator pedal, so that the car still accelerates but at a continuously decreasing rate. If this decrease in acceleration is continued, there comes a time when the marble is undergoing zero acceleration, shown in Fig. 6.1(a) where the straight line intersects the horizontal (time) axis. Thereafter, if the same linear relationship between acceleration and time continues to hold, the acceleration becomes negative, that is, a retardation. Obviously, if the retardation continues long enough, the marble will stop, at time T (say). So to summarize, a marble at rest zero time and zero distance from one edge of the bench starts to move under the influence of a relatively high acceleration. This acceleration declines linearly with time, eventually becoming a retardation until the marble comes to rest again when $t = T$.

Now consider the *distance* the marble travels as a function of time. Evidently while the marble is accelerating, the distance moved in a unit of time will increase, and *vice versa* while the marble is retarding. So the graph of distance moved, s, in time t can be represented by the curve in Fig. 6.1(b). We can assume that the curve shows s as some function of t, i.e. $s = f(t)$, and in order to make further progress, it will be necessary to find a possible form of $f(t)$ which will produce a curve like that in Fig. 6.1(b).

In Fig. 6.2(a) is shown the graph of a third degree polynomial (cf. Fig. 4.3c). The curve is positioned so that: (*i*) it passes through the origin, (*ii*) it rises when t has small positive values, and (*iii*) it ceases to rise beyond a certain value of t (i.e. beyond a certain time) which we call T. Evidently, that part of the curve

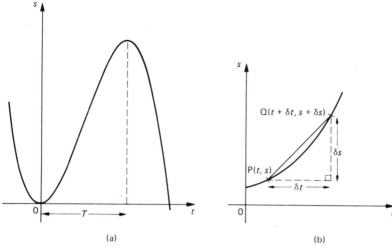

Fig. 6.2   (**a**) A more complete graph of the curve shown in Fig. 6.1(b). (**b**) Two points on a distance–time curve with line PQ depicting average velocity between the points.

lying in the range $0 \leqslant t \leqslant T$ will adequately describe the distance that the marble has travelled, as shown in Fig. 6.1(b). It can be demonstrated that the equation of the curve in Fig. 6.2(a) is

$$s = \tfrac{3}{2}Tt^2 - t^3 \tag{6.1}$$

and if we compare this with the full equation for a third degree polynomial in the same notation, $s = a + bt + ct^2 + dt^3$, we see that the constant term and the term in t are missing (i.e. the coefficients a and b are zero, written $a = b = 0$), $c = \tfrac{3}{2}T$, and $d = -1$. On the other hand, *equation 6.1* only describes the motion of our marble when $0 \leqslant t \leqslant T$. Before zero time, i.e. before the time the marble starts to move, it is considered to be stationary, $s = 0$; the portion of the curve corresponding to negative values of t does not describe the motion of the marble. Similarly, when $t > T$, the marble is at rest distance S from its starting point, whereas the curve of *6.1* shows the marble retracting its steps! Substituting $T = t$ in *6.1* gives $S = \tfrac{1}{2}T^3$. So the full quantitative description of the motion of the marble is given thus:

$$\left. \begin{aligned} s &= \tfrac{3}{2}Tt^2 - t^3 && 0 \leqslant t \leqslant T \\[1em] s &= 0 && t < 0 \\[1em] s &= \tfrac{1}{2}T^3 && t > T \end{aligned} \right\} \tag{6.2}$$

Notice that the variable t does not appear in the last two equations of 6.2, thus showing that s (distance travelled) is independent of time here; i.e. s is constant.

Now imagine two points situated on the distance–time curve (Fig. 6.2b), $P(t, s)$ giving the distance travelled at time t, and $Q(t + \delta t, s + \delta s)$ giving the distance travelled at time $(t + \delta t)$, a short time later. The gradient of the chord PQ is, of course, given by $\delta s / \delta t$. This quotient denotes the amount of distance travelled, δs, in the small interval of time, δt. For a given δt, the larger that δs is, the greater the distance covered. Hence, $\delta s / \delta t$ is a measure of the speed or velocity with which the marble is moving. Moreover, because the marble is undergoing acceleration (Fig. 6.1a), its velocity is not constant; hence $\delta s / \delta t$ is the **average velocity** over the **time interval δt**.

Now we know that the gradient of the curve at P is the limiting value of $\delta s / \delta t$ as $\delta t \rightarrow 0$; we can write this as ds/dt. So as we have identified the gradient of the chord PQ with the average velocity over the time interval δt, then the gradient of the curve (i.e. the gradient of the tangent) at point P must represent a velocity also. Since the time interval is now vanishingly small, there is no question of an average velocity over an interval of time, and the velocity measured by the gradient of the curve at time t is the **instantaneous velocity** of the marble **at time t**.

The idea of an instantaneous velocity may be rather hard to grasp, so consider a slightly different example. When a train is said to be moving at 100 km per hour, this does not mean that it will travel 100 km in the next hour, or that it has travelled 100 km in the previous hour, since its speed will fluctuate. What the statement means is that, 'at the present time the train is travelling at 100 km per hour'. How do we define 'the present time'? The speed of 100 km per hour is equivalent to 27 m per second: we can say that the train is much more likely to travel 27 m in the next second, than 100 km in the next hour, since its speed is much less likely to change in the next second than in the coming hour. It would be even more accurate to say that in the next $\frac{1}{10}$ of a second the train would travel 2.7 m, more accurate still to say that 0.27 m would be traversed in the next $\frac{1}{100}$ of a second, and so on. With a decreasing interval of time, we have a proportionately smaller distance travelled, but nevertheless, the speed of the train is still 100 km per hour, and the statement that the train is travelling at 100 km per hour becomes more accurate as the time interval denoted as 'the present time' becomes smaller. When 'the present time' becomes a vanishingly small interval of time, the statement concerning the speed of the train is as accurate as it can be.

Reverting to our marble moving along a bench, we recall that the distance travelled from its starting point, measured as a function of time, is given by $s = \frac{3}{2}Tt^2 - t^3$. Differentiating,

$$\frac{ds}{dt} = 3Tt - 3t^2 \qquad (6.3)$$

$$\text{or} \quad \frac{ds}{dt} = 3t(T - t) \qquad (6.4)$$

The derivative is also a function of time, and *equation 6.3* shows it to be a second degree polynomial, which has been factorized in *6.4*. Either *equation 6.3* or *6.4*, describes the velocity of the marble when t lies in the range $0 \leqslant t \leqslant T$.

Now when the marble is at rest, velocity is zero, i.e. $ds/dt = 0$. To find the values of t when ds/dt is zero implies solving the quadratic equation

$$3t(T - t) = 0$$

Either $3t = 0$ or $T - t = 0$, i.e. $t = 0$ or $t = T$. This confirms what we already know, that the marble is stationary when $t = 0$ and $t = T$ (Fig. 6.1b).

So to summarize, and put into more general terms, if we have a function of x, such as $y = f(x)$, then $dy/dx = f'(x)$ is the instantaneous velocity of y with respect to x, or the **instantaneous rate of change** of y with respect to x. An **average rate of change** between two specified values of x, x and $x + \delta x$, is given by $\delta y/\delta x$.

Velocity and acceleration – the second derivative

Still using the example of a marble in motion, we have seen that the velocity, for which the symbol v is generally used, is given by ds/dt and bears the following relationship to time:

$$v = \frac{ds}{dt} = 3Tt - 3t^2$$

The curve of this function is shown in Fig. 6.3. We see that, at times 0 and T, v is zero, in other words the marble is at rest at those times. Also, it is clear that the velocity is never constant.

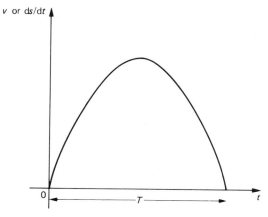

Fig. 6.3   The velocity–time curve, corresponding to the curves in Fig. 6.1 (**a**) and (**b**), whose equation is $v = 3t(T - t)$.

The velocity–time function can, of course, be differentiated, and we have

$$\frac{dv}{dt} = 3T - 6t \qquad (6.5)$$

but since the velocity–time function was obtained from the distance–time function by differentiation, it is also correct to write

$$\frac{d^2 s}{dt^2} = 3T - 6t \qquad (6.6)$$

In other words, the function of t, $3T - 6t$, can be regarded as either the first derivative of the velocity–time function, or the second derivative of the distance–time function describing the motion of the marble. Since dv/dt is a first derivative (of velocity as a function of time), it must denote the rate of change of velocity, i.e. acceleration. If we examine 6.5 we see that it represents a straight line with intercept $3T$ and a slope of -6; it is, in fact, the acceleration–time relationship that we postulated at the beginning of this discussion (Fig. 6.1a).

Bearing in mind that a velocity is a rate of change of one variable with respect to another, and that acceleration is rate of change of velocity, this means that, in general terms, a second derivative must be the ***rate of change of the rate of change*** of a function with respect to its variable. In symbols, if $y = f(x)$, then $dy/dx = f'(x)$ is the rate of change of y with respect to x, and $d^2 y/dx^2 = f''(x)$ is the rate of change of the rate of change of y with respect to x. The latter is also called the ***acceleration*** of y with respect to x; if $d^2 y/dx^2$ is a negative quantity, it is often called 'retardation'. To facilitate comparisons, the distance–time, velocity–time and acceleration–time graphs are all shown on a common time-axis in Fig. 6.4.

The growth of a microbial culture

Cell number as a function of time

We shall now begin discussing a biological phenomenon to which some of the concepts of the differential calculus can be applied.

Suppose we inoculated a known volume of a suitable nutrient medium with a species of bacterium. After mixing to ensure a homogeneous dispersion of the bacterial cells, a known small volume of the suspension is sampled and the number of cells contained therein is estimated. By knowing the volume of the sample and the volume of the nutrient medium in which the inoculum was dispersed, it is easy to obtain an estimate of the total number of cells in the inoculum; give this number the symbol n_0. Now incubate the culture and periodically sample it, as described above, so as to obtain numbers of bacterial

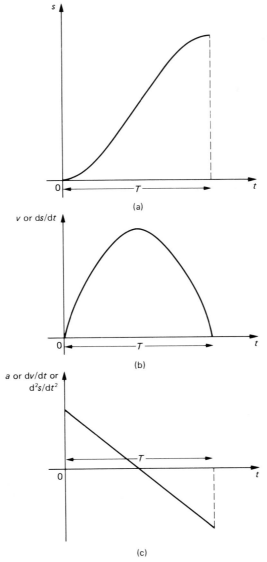

Fig. 6.4 The distance–time, velocity–time, and acceleration–time curves of Figs 6.1 and 6.3 shown with the same time scale.

cells, n, at a series of times, t. These observations can be plotted on a graph, where $t = 0$ is the time of inoculation. This graph will normally take a form similar to that shown in Fig. 6.5(a). The fundamental question then arises, 'why does the growth of the culture take this particular form?'

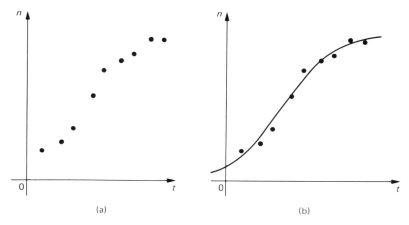

Fig. 6.5   (**a**) An hypothetical set of data showing the growth of a microbial culture. (**b**) The same, but with a fitted function.

Now it is reasonable to suppose that if we had taken many more samples at closer intervals of time, the extra points on the graph would have been situated between the points that are present on Fig. 6.5(a) with respect to both times and numbers. Taking a small portion of the total volume is called ***sampling***, and the small volume actually taken is a ***sample***.

Consider a particular sampling time. The actual sample taken will contain a number of bacterial cells, indicated by the position of the point on the graph. But if at this particular time a different portion of the culture (different sample) had been taken, it is very unlikely that the number of bacteria counted would have been identical with the number in the actual sample. At this particular time a whole series of samples could be taken, and each would indicate a slightly different value of n. Variation of this kind is evidently bound up with the process of sampling, and has nothing to do with the complete entity (bacterial culture in this case); so the name ***sampling variability***, or sampling error, is given to this phenomenon. A common practice, however, is to take only a single sample at any one time, and the assumption is made that, aside from sampling error, the points on the graph would lie on a smooth curve, as shown in Fig. 6.5(b), which we can infer has an underlying functional relationship of the form $n = f(t)$. The question which now arises is, 'what is the form of $f(t)$?'

A growth model based on rates of change – differential equations

There are a number of functions which, from an empirical point of view, would provide good descriptions of the increase of cell number with time in our culture solution; but arbitrarily selecting one of these equations would not help us to elucidate the underlying mechanisms of growth of the bacterial culture.

Instead, we attempt to construct a simple mathematical model of growth, by marshalling together relevant facts that we think are already established and putting them into quantitative form. For many biological situations, the present one included, it is natural to formulate the model in terms of rates of change.

This kind of approach, involving the construction of a mathematical model, was referred to in Chapter 1. It was stated that the results produced by the model may or may not agree with experimental findings; if they do not it means that one or more of the assumptions made in constructing the model are untenable. In the following model of the growth of a bacterial culture, although the assumptions made appear to be plausible, the final result of the model, a function of the form $n = f(t)$, does not quite agree with growth curves obtained by careful experimental work. This, however, is not important for our present purpose, which is to show how a mathematical model of a biological process may be constructed using the concepts of the differential calculus. The model is broadly correct and, as will be seen, over-simplification in one of the assumptions is the cause of the discrepancy.

With regard to the multiplication of bacterial cells, it is known that in a constant but unlimited environment each cell, on average, divides at a constant rate. Hence the more cells there are present, the more rapid will be the rate of increase for the entire colony. Hence, we can write

$$\frac{dn}{dt} \propto n \qquad (6.7)$$

i.e. the rate of increase of cell number at an instant of time is proportional to the number of cells present at that time. The proportionality sign can be changed for an 'equals' if a constant of proportionality is introduced on the right-hand side of 6.7:

$$\frac{dn}{dt} = kn \qquad (6.8)$$

Notice that: (*i*) 6.8 takes the form $dn/dt = f(n)$ rather than $dn/dt = f(t)$: the rate of change of cell number is a function of cell number, not time; (*ii*) in (*i*) we have written $dn/dt = f(n)$ rather than $dn/dt = f'(n)$, since the latter would imply that we had started off with $f(n)$ and differentiated it to $f'(n)$.

Equation 6.8 represents a straight line on a graph whose horizontal axis is n, and whose vertical axis is dn/dt. The line passes through the origin (intercept of zero), has a gradient of k, and is shown in Fig. 6.6(a). Evidently, the higher the value of k, the greater the rate of growth of the culture for a given number of cells, n. Notice also that the relationship is valid at the origin, which means that when no organisms are present, the rate of increase is zero – a very sensible conclusion! In fact, the original inoculum would contain a relatively small number of cells, n_0, which, substituted into 6.8, gives a relatively small growth rate for the culture. The gradient, k, is specific to a given species of bacterium under defined environmental conditions.

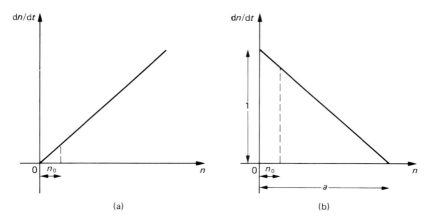

Fig. 6.6   (**a**) The straight line $dn/dt = kn$. (**b**) The straight line $dn/dt = 1 - n/a$.

Now, of course, there is no such thing as an unlimited environment, because this would mean that one bacterial cell would have infinite space and resources at its disposal. For this example, however, we can say that at the time of inoculation the environment is *relatively* unlimited, and as the organisms multiply they use up the nutrients and secrete waste products, some of the latter being actually toxic to the cells. Hence, we can postulate that, as the culture grows, the rate of increase is declining; moreover, it is evident that the decline in growth rate must be some function of the (increasing) number of cells,

$$\text{i.e.} \qquad \frac{dn}{dt} = f(n) \qquad\qquad (6.9)$$

What form can $f(n)$ take to comply with our theoretical requirements? The simplest form of $f(n)$ that is suitable is shown graphically in Fig. 6.6(b) – a straight line with a negative slope.

Equation 6.8 is a quantitative statement of an intrinsic property of the microbial culture, while *equation 6.9* represents the sum total of increasing environmental restrictions. The actual growth rate of the culture at any instant will be the resultant of both processes represented by *6.8* and *6.9*, so we can write

$$\frac{dn}{dt} = knf(n) \qquad\qquad (6.10)$$

that is, the growth rate of the culture at any instant is the product of the two processes. We have said that $f(n)$ is a linear function, but we have not yet specified values for its slope and intercept. When n is very small (say 1), the environment is virtually unlimited, and if n could be zero, then the environment would be truly unlimited. Hence, when n is very small, dn/dt is essentially given by *6.8*, and this means that $f(n)$ in *6.9* and *6.10* at this time must be approximately unity. So the intercept of the straight line can be given the value of 1. Quantification of the gradient is rather more arbitrary. Let us specify the degree of tolerance to declining nutrients and accumulating wastes as a; then if a is large (i.e. there is a high tolerance) the decline of dn/dt with increasing n is slow, and this is associated with a small negative slope of the straight line. Conversely, if the organisms show only a low tolerance to an increasingly adverse environment, a is small, and the gradient of the line is steep. Evidently, the gradient can be specified as $-1/a$, and so *6.9* becomes

$$\frac{dn}{dt} = 1 - \frac{1}{a} n \qquad (6.11)$$

and *6.10* becomes

$$\frac{dn}{dt} = kn \left(1 - \frac{n}{a} \right) \qquad (6.12)$$

Looking first at *6.11*, if we put $n = a$, $dn/dt = 0$. Hence, the line intersects the horizontal axis at a distance of a from the origin. At the point of intersection dn/dt is zero, and the growth rate has been slowing down as $n \rightarrow a$. This means that $n = a$ is the greatest number of cells that can occur in this culture under these defined conditions, and so the constant a has a second meaning in relation to the growth of a bacterial culture.

Turning our attention now to *equation 6.12*, which quantifies the overall relationship between rate of growth and culture density, multiply out the bracket:

$$\frac{dn}{dt} = kn - \frac{k}{a} n^2 \qquad (6.13)$$

In this form, it is evident that $dn/dt = f(n)$ is a second degree polynomial. Referring to the general equation of a second degree polynomial (page 45), we see that the constant term is missing in *6.13*, and hence the curve passes through the origin; also, when $n = a$ in *6.12* or *6.13*, dn/dt is zero. The coefficient of n^2 is negative, hence the curve lies as shown in Fig. 6.7 (cf. Fig. 4.1b). The rate of growth of the culture rises at first while the environment is not very restrictive; loosely, we can say that *6.8* is mainly controlling growth rate during this phase, and the resistance of *6.11* is small. However, the effect of the latter is increasing, while the effect of the former stays constant. So the overall growth rate rises to a maximum and then declimes, as *function 6.11* approaches zero and thus exerts an ever-increasing effect over the overall growth rate, as described by *6.12*.

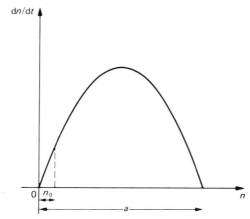

Fig. 6.7   Curve of the second degree polynomial $dn/dt = kn(1 - n/a)$.

Equations such as *6.8*, *6.11*, and *6.12*, are known as **differential equations**, because they contain a differential coefficient; so in general a **differential equation describes a rate of change.** The differential equations of this example are of the form $dn/dt = \phi(n)$; that is, the instantaneous rate of growth of the culture is a function of culture density at that instant. We commenced this discussion by looking for a function of the form $n = f(t)$, which would tell us the number of cells in the culture at any given time. Now it is possible to obtain $n = f(t)$ from $dn/dt = \phi(n)$, and such a procedure is referred to as 'solving the differential equation'.

Differential equations and their solution form a very important, and difficult, part of mathematics. We are not yet in a position to solve even the simplest equations that have arisen in the preceding discussion. The core procedure in the solution of a differential equation is the reverse process of differentiation, namely, integration; but even when we have considered integration and its methods in the next chapter, there are other topics to be dealt with before we are able to solve an equation like *6.12*, and we shall return to the topic later in the book (Chapter 8).

Two final points must be made about the model of bacterial culture growth that has been described. In Fig. 6.5(b), the culture has been assumed to grow according to a smooth curve, implying that cell number can take fractional values, which is clearly wrong. In fact, the actual form of the curve is a series of steps, each step corresponding to a cell division; but the curve shown in Fig. 6.5(b) would go through the series of steps, and is merely an approximation (Fig. 6.8). If the number of organisms was relatively few, such as in a population of higher plants or animals, the curve would be only a gross approximation; but in the case of the microbial colony, which involves thousands or millions of individuals, the approximation is good.

Secondly, it was indicated at the beginning of this discussion (page 91) that the result of this model, which is a function of the form $n = f(t)$, does not accord with experimental observation. This means that the curve of the

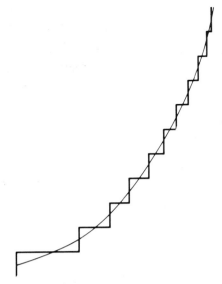

Fig. 6.8   A smooth curve approximating a 'step function'.

function $n = f(t)$ is somewhat dissimilar to a curve produced from an actual growth experiment in which sampling is frequent. Although the curve produced from an experiment is broadly sigmoid in form (as is the curve of the function $n = f(t)$ derived from the differential equation *6.12*) close inspection shows it to have three distinct phases: a lag phase at the beginning where very little increase in cell number occurs, a relatively prolonged phase of growth in which the rate of increase of cell number is proportional to the number present (as described by *equation 6.8*), and only a short phase of growth rate decline. Evidently, our first assumption embodied in *equation 6.8* is satisfactory; but our second is open to question. Broadly, *equation 6.9* is correct, but to interpret $f(n)$ as a simple linear function as shown in *equation 6.11* is wrong, and a search must be made for another, more realistic and more complex, form of $f(n)$ which gives a result for the model that accords better with experimental findings (see Exercise 1 at the end of this chapter).

Maxima and minima

Maximum growth rate of the microbial culture

In the previous discussion of the growth of a bacterial culture we saw that the growth rate increased while the number of individuals was small, but decreased when the number of organisms was relatively high. This means that the rate of increase must rise to some maximum value before the decline in growth rate ensues (Fig. 6.7). Two questions may be asked: (*i*) what is the maximum growth rate, and (*ii*) how dense is the culture when its growth rate is at the maximum? Assuming we knew the values of the constants, *a* and *k*, approximate answers to these questions could be obtained by drawing a graph. But it is much quicker, and more accurate, to utilize the differential calculus for this purpose. We shall do this for our microbial culture in this section, and then proceed to a more general discussion of maximum and minimum points.

Referring to Fig. 6.7, it can be seen that the gradient of the curve, that is the rate of change of the growth rate (acceleration), is positive when *n* is small, and negative when *n* is large after the maximum growth rate. As *n* approaches the point where the maximum growth rate occurs, the gradient gets smaller and smaller. At the point of maximum growth rate, the gradient of the curve is zero, and thereafter it is increasingly negative. The essential point of the last two sentences is that the gradient of the curve is zero when dn/dt is at its maximum, and since in general terms the curve represents a function of the form

$$\frac{dn}{dt} = f(n)$$

which differentiated with respect to *n* gives
$$\frac{d(dn/dt)}{dt} = f'(n)$$

this implies that $f'(n) = 0$

at the maximum point. Differentiating *6.13* with respect to *n*:

$$\frac{d(dn/dt)}{dn} = k - \frac{2k}{a}n \qquad (6.14)$$

and on equating the right-hand side of *6.14* to zero, we have

$$k - \frac{2k}{a}n = 0$$

i.e. $k(1 - 2n/a) = 0$, which yields the result $n = a/2$. Hence the maximum growth rate occurs when the colony is half-grown. To find what the maximum rate of growth actually is, we substitute $a/2$ for *n* in *6.13*.

This gives
$$\frac{dn}{dt} = k\left(\frac{a}{2}\right) - \frac{k}{a}\left(\frac{a}{2}\right)^2$$

$$= \frac{ka}{2} - \frac{ka^2}{4a} = \frac{ka}{2} - \frac{ka}{4}$$

i.e.
$$\frac{dn}{dt} = \frac{ka}{4}$$

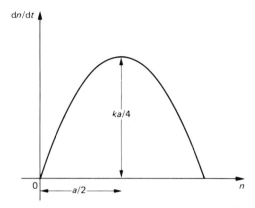

Fig. 6.9 Curve of the second degree polynomial $dn/dt = kn(1 - n/a)$, showing the maximum value of dn/dt, and the value of n at which the maximum occurs.

The curve of growth rate against cell number is reproduced in Fig. 6.9, showing the quantities just calculated.

Maximum and minimum, general

Consider the curve in Fig. 6.10, which depicts a function $y = f(x)$ that has several maximum and minimum points. It is important to note exactly what constitutes a maximum or a minimum point (on a curve). A ***maximum point*** is one which has a greater y-value than any *adjacent points* on the curve; hence A and C are *both* maximum points. A ***minimum point*** is one which has a lower y-value than any adjacent points on the curve; thus B and D are both minimum points. Evidently, maximum and minimum points are 'local phenomena', and

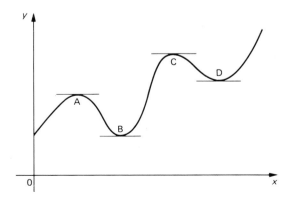

Fig. 6.10 Maxima and minima on a curve.

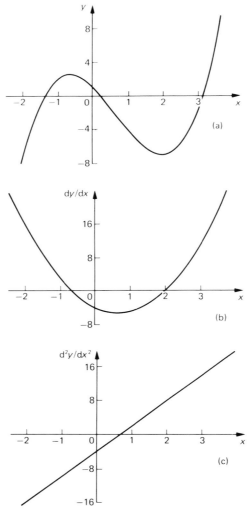

Fig. 6.11 (**a**) Curve of the function $y = x^3 - 2x^2 - 4x + 1$. (**b**) Curve of the function $dy/dx = 3x^2 - 4x - 4$. (**c**) The straight line $d^2y/dx^2 = 6x - 4$.

each is defined for its own part of the curve. There is no question of a maximum point being necessarily the highest y-value the curve can attain anywhere, although in many instances this may happen to be true; and conversely for a minimum point. In Fig. 6.10, the minimum point D has a greater y-value than the maximum point A; but in Fig. 6.9, in which the curve is a second degree polynomial, there is no minimum point and the maximum point does indeed represent the highest value that the dependent variable can attain.

Maxima and minima are collectively known as **turning points**, because the gradient of the curve changes from positive to negative or *vice versa* at these locations.

Example 6.1

Find the turning points on the curve of the function

$$y = x^3 - 2x^2 - 4x + 1 \qquad (6.15)$$

This is a cubic polynomial, and evidently there is one maximum and one minimum point (Fig. 6.11a). To find the positions of these points, we differentiate the function and equate the derivative to zero

$$\frac{dy}{dx} = 3x^2 - 4x - 4$$

hence $3x^2 - 4x - 4 = 0 \qquad (6.16)$

The *relationship 6.16* is a quadratic equation which can be solved by the *formula 4.2* applied to the quadratic *equation 4.1*. In the present instance, we have

$$x = \frac{4 \pm \sqrt{\{(-4)^2 + 48\}}}{6}$$

$$= \frac{4 \pm \sqrt{64}}{6} = \frac{4 \pm 8}{6}$$

Hence $x = 2$ or $x = -\frac{2}{3}$

Substituting these x-values successively back into the original *equation 6.15*

$$y = (2)^3 - 2(2)^2 - 4(2) + 1$$

$$= 8 - 8 - 8 + 1 = -7$$

and $y = (-\frac{2}{3})^3 - 2(-\frac{2}{3})^2 - 4(-\frac{2}{3}) + 1$

$$= -\frac{8}{27} - \frac{8}{9} + \frac{8}{3} + 1$$

$$= \frac{-8 - 24 + 72 + 27}{27} = \frac{67}{27}$$

$$= 2\frac{13}{27}$$

So the points $(2, -7)$ and $(-\frac{2}{3}, 2\frac{13}{27})$ are turning points of the curve of $y = x^3 - 2x^2 - 4x + 1$.

To determine which of the turning points is a maximum and which is a minimum, without looking at the graph, it is necessary to consider the second derivative. On viewing Fig. 6.11(a) again, we see that around a maximum point the gradient is at first positive but becomes less so as the maximum point is approached; is zero at the point, and then becomes increasingly negative when the maximum point is passed. In other words, the gradient is continually declining in absolute value through a maximum point, which means that the rate of change of the gradient $(d^2 y/dx^2)$ is negative. The converse applies at a minimum point where $d^2 y/dx^2$ is positive. The determination of whether a particular turning point is a maximum or a minimum, then, resolves itself into examining the value of the second-order differential coefficient at that point.

Example 6.2

Determine the nature of the turning points of the function in *example 6.1*.

$$\text{We have} \qquad y = x^3 - 2x^2 - 4x + 1$$

$$\text{and} \qquad \frac{dy}{dx} = 3x^2 - 4x - 4$$

$$\text{hence} \qquad \frac{d^2 y}{dx^2} = 6x - 4 \qquad\qquad (6.17)$$

The two turning points are at $x = 2$ and $x = -\frac{2}{3}$, so substituting successively for x into *6.17*, we have

$$\text{for } x = 2 \qquad \frac{d^2 y}{dx^2} = 6(2) - 4 = 8$$

$$\text{for } x = -\frac{2}{3} \qquad \frac{d^2 y}{dx^2} = 6(-\frac{2}{3}) - 4 = -8$$

When $x = 2$, $d^2 y/dx^2$ is a positive quantity, and so the point $(2, -7)$ is a minimum. When $x = -\frac{2}{3}$, $d^2 y/dx^2$ is a negative quantity, and so the point $(-\frac{2}{3}, 2\frac{13}{27})$ is a maximum. These statements can be checked by reference to Fig. 6.11(a).

The relationships between the functions $y = f(x)$, $dy/dx = f'(x)$, and $d^2 y/dx^2 = f''(x)$ for the above example are illustrated in Fig. 6.11. In Fig. 6.11(a) is shown the curve of the original function $y = f(x)$, and in Fig. 6.11(b) is shown the curve of the function giving the first derivative of $f(x)$ on the same scale of x. Thus for any value of x, the ordinate in Fig. 6.11(b) is the actual value of the gradient of the curve in Fig. 6.11(a) at that x-value. As already remarked, the gradient of $y = f(x)$ at a turning point is zero, and in this example the turning points are at $x = 2$ and $x = -\frac{2}{3}$. Thus we see in Fig. 6.11(b) that the curve of dy/dx intersects the horizontal axis at these two values of x. Working from left to right along the x-axis, before the maximum point the gradient of $y = f(x)$ is positive, and so the curve of dy/dx is above the x-axis. Similarly, after the minimum point on $y = f(x)$ the gradient is likewise positive, and so the curve of dy/dx is again above the x-axis. Between the maximum and minimum points, however, the gradient of $y = f(x)$ is negative, and so over this range of x, i.e. $-\frac{2}{3} < x < 2$, the curve representing dy/dx is below the x-axis.

Points of inflexion

Look again at Fig. 6.11(a) and recollect how the gradient of a curve changes around a maximum and around a minimum point. We have seen that around a maximum point the rate of change of gradient is negative (Fig. 6.11b); after the maximum point is passed, the gradient itself is negative and becomes more so. Before the minimum point, however, we have seen that although the gradient is still negative, it becomes less so as the minimum point is approached (Fig. 6.11b). Evidently there must be some point between the maximum and the minimum where the gradient stops becoming more and more negative, and commences to be less negative as one moves along the x-axis from left to right. Such a point is known as a point of inflexion and, unlike maxima and minima, it is not possible to obtain a reasonable approximation of the position of a point of inflexion by graphical methods; the calculus must be used.

We have just said that at a point of inflexion the gradient of the curve ceases becoming more negative and starts to become less negative (i.e. more positive). While the gradient is becoming more negative, the rate of change of gradient, $d^2 y/dx^2$, is negative. When the gradient is becoming less negative, i.e. more positive, $d^2 y/dx^2$ is positive. So at the point of inflexion, $d^2 y/dx^2$ changes from being a negative quantity to a positive one, i.e. $d^2 y/dx^2 = 0$ (Fig. 6.11c). This discussion is based on the example illustrated in Fig. 6.11, where the maximum of the curve precedes the minimum. If a minimum preceded a maximum, the above comments describing the rate of change of the gradient would be reversed, but nevertheless, the key property of a zero second derivative at the point of inflexion would remain.

Example 6.3
Determine the co-ordinates of the point of inflexion of the function employed in *examples 6.1* and *6.2*. We have, *6.17*,

$$\frac{d^2 y}{dx^2} = 6x - 4$$

Equating this to zero, gives $x = \frac{2}{3}$. Substituting back in the original function,

$$y = \frac{8}{27} - \frac{8}{9} - \frac{8}{3} + 1 = -2\frac{7}{27}$$

Hence, the point of inflexion is $(\frac{2}{3}, -2\frac{7}{27})$.

The relationship between d^2y/dx^2, dy/dx, and $y\{=f(x)\}$ can be appreciated by comparing Fig. 6.11(c) with (b) and (a).

EXERCISES

1. At the end of the section of this chapter describing the growth of a microbial culture, it was pointed out that while the assumption embodied in *equation 6.8* was satisfactory, the assumption quantified in *equation 6.11* was untenable. Consider the following three model equations:

$$\frac{dn}{dt} = 0.2n \tag{6.20}$$

$$\frac{dn}{dt} = 1 - \frac{n}{1\ 000\ 000} \tag{6.21}$$

$$\frac{dn}{dt} = 1 - \left(\frac{n}{1\ 000\ 000}\right)^2 \tag{6.22}$$

and draw graphs of these relationships in the range $0 \leqslant n \leqslant 1\ 000\ 000$. Combine *equations 6.20* and *6.21* to give the complete growth model (cf. *equations 6.12* and *6.13*). Calculate the co-ordinates of the maximum point, and draw the graph of the complete growth model.

Similarly, combine *equations 6.20* and *6.22* to form a complete growth model; calculate the co-ordinates of the maximum point, and draw the graph on the same axes as the previous model. What are your conclusions?

2. Find the co-ordinates of the maximum and minimum points, and of the point of inflexion of the function $f(x) = x^3 - 3x + 1$.

7

Integration

In this chapter we shall study the other branch of calculus, whose core feature is the process of integration. In its basic form, integration is the converse of differentiation, and we examine this aspect of integration first. It has already been mentioned (Chapter 5, page 64) that integration can also be regarded as a summation process, and we shall deal with this equally important aspect of the subject later in this chapter.

Basic concepts – the indefinite integral

Consider a function of x, say x^3. Then we know that

$$\frac{d(x^3)}{dx} = 3x^2 \qquad (7.1)$$

that is, $3x^2$ is the first differential coefficient or first derivative of x^3. Reversing the notation, we say that x^3 is the *integral* of $3x^2$, and we write

$$\int 3x^2 \, dx = x^3 \qquad (7.2)$$

In more general terms, we have written

$$\frac{d\{F(x)\}}{dx} = F'(x) \qquad (7.3)$$

(page 75). The symbol $F'(x)$ implies that, starting from a known function $F(x)$, its derivative has been found. Now in situations in which integration is used, we only know the function $F'(x)$, and required to find $F(x)$. Hence, it is better to give $F'(x)$ a different notation, $f(x)$; then 7.3 becomes

$$\frac{d\{F(x)\}}{dx} = f(x) \qquad (7.4)$$

Reversing *7.4*, cf. *7.1* and *7.2*, we have that

$$\int f(x)\,\mathrm{d}x = F(x) \qquad\qquad (7.5)$$

On the left-hand side of *relationship 7.5*, $\int$ is the **integral sign**, $f(x)$ is known as the **integrand**, and $\mathrm{d}x$ is the **variable of integration**; one can say that $f(x)$ is **integrated with respect to** x. The whole expression $\int f(x)\,\mathrm{d}x$ is called an **integral.**

A fundamental difference between the processes of differentiation and integration is that the former can be appreciated and carried out from first principles, whereas the latter cannot be done from first principles. Integration is purely the converse of differentiation: while rules exist to enable any function to be differentiated, the same is not true of integration. We often run into serious problems when attempting to integrate even quite simple-looking functions.

The indefinite integral and the constant of integration

Let $y = x^3$; then $\mathrm{d}y/\mathrm{d}x = 3x^2$. But suppose, instead, that $y = x^3 + 4$; still we find $\mathrm{d}y/\mathrm{d}x = 3x^2$. To be quite general, suppose that $y = x^3 + c$, where c is any constant, again $\mathrm{d}y/\mathrm{d}x = 3x^2$. Since integration is the converse of differentiation, we may therefore write

$$\int 3x^2\,\mathrm{d}x = x^3 \qquad \text{in the first case}$$

$$\text{or} \quad \int 3x^2\,\mathrm{d}x = x^3 + 4 \qquad \text{in the second case}$$

$$\text{or} \quad \int 3x^2\,\mathrm{d}x = x^3 + c \qquad \text{in general;}$$

and c may be *any* constant. This apparent anomaly arises because the derivative of a constant is zero. So if we are given $\mathrm{d}y/\mathrm{d}x = 3x^2$ and are asked to say what y is equal to, we have no way of knowing whether $y = x^3$, or $y = x^3 + 4$, or $y = x^3 + c$ where c is any constant number. To allow for this contingency, we write the result of an integration problem such as

$$y = \int 3x^2\,\mathrm{d}x$$

$$\text{as} \quad y = x^3 + c$$

where c is known as the **constant of integration**. Typically, the value of c is unknown to us, and consequently $\int 3x^2 \, dx$ is an example of an **indefinite integral**. In more general terms, wherever integration is used as the converse of differentiation, it is the indefinite integral that is involved; the constant of integration must be allowed for, but it is usually unknown.

Equations 7.4 and *7.5* together serve to define the nature of integration, and can be called 'the principle of integration'. The evaluation of a derivative is a direct operation for which precise rules exist to obtain a unique result. No such rules exist for integration, and in the last resort the function to be integrated must be recognizable as the result of a differentiation of some kind. A check that an integration has been correctly performed is to differentiate the function obtained as a result of the integration, and to ensure that the differentiated function is equal to the original function.

Integration is not, however, entirely the hit-and-miss process implied in the previous paragraph. There are certain results, which are called **standard integrals**, and many methods by which particular functions can be converted to standard integrals, and so integrated. On the other hand, although any function can be differentiated, there are many functions that cannot be integrated; we shall return to this problem later.

Integration of simple functions

The integration of polynomial functions

If $f(x) = x^m/m$, where m is a constant, we know that $f'(x) = mx^{(m-1)}/m = x^{(m-1)}$. In other words, $\int x^{(m-1)} \, dx = x^m/m + c$. Integration has been carried out by adding one to the power of x, that is $(m - 1) + 1$, and dividing by the new power, m.

Thus we have the standard integral

$$\int x^n \, dx = \frac{x^{(n+1)}}{n+1} + c \qquad (7.6)$$

where n is any real number, except -1. (If n were -1, we should have to divide x^0 by 0, which gives no sensible result.)

Example 7.1
Find

$$(a) \int x^2 \, dx \quad (b) \int dx/x^2 \quad (c) \int \sqrt{x} \, dx$$

(a) We have immediately from *7.6*

$$\int x^2 \, dx = x^3/3 + c$$

(b)
$$\int dx/x^2 = \int \frac{1}{x^2} dx = \int x^{-2} dx$$

Hence, on applying 7.6
$$\int x^{-2} dx = \frac{x^{(-2+1)}}{-2+1} + c$$

Thus
$$\int dx/x^2 = -1/x + c$$

(c)
$$\int \sqrt{x}\, dx = \int x^{1/2} dx = \frac{x^{(1/2+1)}}{\frac{1}{2}+1} + c = \frac{x^{3/2}}{\frac{3}{2}} + c$$

Therefore
$$\int \sqrt{x}\, dx = \tfrac{2}{3}\sqrt{x^3} + c$$

In differentiation, we found that if we had a function of x which was made up of simpler functions of x added to, or subtracted from, one another, then differentiation of the whole function was carried out by dealing with each term separately;

i.e. if $f(x) = \phi(x) \pm \psi(x)$

then $f'(x) = \phi'(x) \pm \psi'(x)$

Since integration is the converse of differentiation, the same rule will apply in integration;

hence if $f(x) = \phi(x) \pm \psi(x)$

then $\int f(x)\, dx = \int \phi(x)\, dx \pm \int \psi(x)\, dx + c$ (7.7)

Also in differentiation, if $f(x) = ax^n$, then $f'(x) = anx^{(n-1)}$, the coefficient a going through the process unchanged; the same is true for a coefficient in integration, and we may therefore place such a constant outside the integral sign:

$$\int ax^n\, dx = a \int x^n\, dx = \frac{ax^{(n+1)}}{n+1} + c$$ (7.8)

The centre term in 7.8 consists of a multiplied by the standard integral on the left-hand side of 7.6, so the result of integrating ax^n with respect to x is readily obtained.

Example 7.2
   Find

$$\text{(a)} \ \int 2x^3 \, dx \quad \text{(b)} \ \int \frac{x \, dx}{2} \quad \text{(c)} \ \int \frac{5\sqrt{x} \, dx}{3}$$

(a)

$$\int 2x^3 \, dx = 2 \int x^3 \, dx = 2x^4/4 + c$$

   Therefore $\int 2x^3 \, dx = \tfrac{1}{2}x^4 + c$

(b)

$$\int \frac{x \, dx}{2} = \tfrac{1}{2} \int x \, dx = \tfrac{1}{2} \frac{x^2}{2} + c$$

   Hence $\int \frac{x \, dx}{2} = \tfrac{1}{4}x^2 + c$

(c)

$$\int \frac{5\sqrt{x} \, dx}{3} = \tfrac{5}{3} \int x^{1/2} \, dx = \tfrac{5}{3} \frac{2x^{3/2}}{3} + c$$

   Thus $\int \frac{5\sqrt{x} \, dx}{3} = \tfrac{10}{9} \sqrt{x^3} + c$

The integration of $(a + bx)^n$

Another useful result is

$$\int (a + bx)^n \, dx = \frac{(a + bx)^{(n+1)}}{b(n + 1)} + c, \quad n \neq -1 \qquad (7.9)$$

The standard integral *7.9* is similar to that of *expression 7.6*, and is valid for any value of n except -1. No derivation of the result in *7.9* is given here, because it is simple enough to differentiate the right-hand side (by the function of a function rule) and see that this does yield the left-hand side. Indeed, as we have already remarked (page 105), integration may often have to be attempted on such a trial and error basis.

Partial fractions and their use in integration

As already remarked, the process of integration of a function requires first that the function be resolved into a form recognizable as a standard integral. There are several methods, of varying complexity, for attempting this process;

hitherto we have only used the trivial ideas of placing a coefficient outside the integral sign, and resolving the integral of a polynomial into a sum of integrals of each individual term.

We shall now examine a method of greater complexity for changing certain intractable-looking integrals into standard ones.

If we are given the expression

$$\frac{2}{3-x} + \frac{1}{2+x}$$

it is easy to amalgamate the two fractions, i.e. adding them together by putting them over a common denominator:

$$\frac{2(2+x) + (3-x)}{(3-x)(2+x)} = \frac{4 + 2x + 3 - x}{(3-x)(2+x)} = \frac{7+x}{(3-x)(2+x)}$$

Thus $\quad \dfrac{2}{3-x} + \dfrac{1}{2+x} = \dfrac{7+x}{(3-x)(2+x)}$ $\qquad$ (7.10)

So, working from left to right in *7.10* is easy.

Now the two fractions on the left-hand side can be recognized as standard integrals (*equation 9.28* on page 159):

$$\int \frac{2\,dx}{3-x} = 2\int \frac{dx}{3-x} = -2\log_e (3-x) + c$$

and $\quad \displaystyle\int \frac{dx}{2+x} = \log_e (2+x) + c$

but the right-hand side of *7.10* cannot be recognized as a standard integral as it stands. Hence, if we were required to find

$$\int \frac{7+x}{(3-x)(2+x)}\,dx$$

it would first of all be necessary to decompose this fraction into its **partial fractions**; in other words, we require a process whereby we may work from right to left in *7.10*. Such a method is known simply as 'partial fractions'.

Returning to the above example, we first of all know that the partial fractions of $(7+x)/\{(3-x)(2+x)\}$ will be two in number, and one will have $(3-x)$ in its denominator and the other will have $(2+x)$ in its denominator.

The numerator of each partial fraction will be unknown: let us represent these as A in one and as B in the other partial fraction. Then we can write

$$\frac{7 + x}{(3 - x)(2 + x)} \equiv \frac{A}{3 - x} + \frac{B}{2 + x} \qquad (7.11)$$

The $\equiv$ sign means that *relationship 7.11* is an identity, i.e. that the right-hand and left-hand signs must be equal to one another for *all* values of the variable x. (An $=$ sign does not imply this: the equality may hold for only certain values of x.)

Now, multiply both sides of *7.11* by $(3 - x)(2 + x)$, the denominator of the fraction whose partial fractions are required, and we have

$$7 + x \equiv A(2 + x) + B(3 - x) \qquad (7.12)$$

Since *relationship 7.12* must hold for all values of x, A and B may be determined by selecting appropriate values for x and substituting them successively into *7.12*. First put $x = 3$, when it is obvious that the second term on the right-hand side of *7.12* disappears; this leaves only A as an unknown and we may solve for it. Then put $x = -2$, which in a similar fashion leaves only B as an unknown which can be solved for.

So, put $\quad x = 3$: $\quad$ then $\quad 7 + 3 = A(2 + 3)$
$\qquad\qquad\qquad\qquad$ i.e. $\qquad 10 = 5A \qquad$ and so $\qquad A = 2.$

Put $\quad x = -2 \quad$ then $\quad 7 + (-2) = B\{3 - (-2)\}$
$\qquad\qquad\qquad\qquad$ i.e. $\qquad 5 = 5B \qquad$ and so $\qquad B = 1.$

Hence the two partial fractions are, from *7.11*, $2/(3 - x)$ and $1/(2 + x)$, which we see from *7.10* are correct. Thus, we have that

$$\int \frac{(7 + x)\,dx}{(3 - x)(2 + x)} = \int \frac{2dx}{3 - x} + \int \frac{dx}{2 + x} = \log_e (2 + x) - 2 \log_e (3 - x) + c$$

Example 7.3
  Integrate

$(a) \quad \dfrac{3x^2 - 7}{x(x - 2)(x + 4)}$ $\qquad\qquad (b) \quad \dfrac{x}{x^2 + 5x + 6}$

(a) First, the expression is decomposed into partial fractions.

$$\text{Let} \qquad \frac{3x^2 - 7}{x(x-2)(x+4)} \equiv \frac{A}{x} + \frac{B}{x-2} + \frac{C}{x+4}$$

(The fact that the numerator of the composite fraction contains a function of x does not affect the method so long as the highest power of x in the numerator is lower than the highest power in the denominator.) Multiply both sides by $x(x-2)(x+4)$:

$$3x^2 - 7 \equiv A\,(x-2)(x+4) + Bx(x+4) + Cx(x-2)$$

Put $x = 2$:   then $3(2)^2 - 7 = B(2)(6)$ i.e. $12 - 7 = 12B$

                             so $B = \frac{5}{12}$

Put $x = -4$:   then $3(-4)^2 - 7 = C(-4)(-6)$ i.e. $48 - 7 = 24C$

                             so $C = \frac{41}{24}$

Put $x = 0$:   then $-7 = A(-2)(4)$ i.e. $-7 = -8A$

                             so $A = \frac{7}{8}$

$$\text{Hence} \qquad \frac{3x^2 - 7}{x(x-2)(x+4)} = \frac{7}{8x} + \frac{5}{12(x-2)} + \frac{41}{24(x+4)}$$

$$\text{Now} \qquad \int \frac{7\,dx}{8x} = \tfrac{7}{8} \int \frac{dx}{x} = \tfrac{7}{8} \log_e x + c \quad \text{(page 157)}$$

$$\left. \begin{array}{l} \int \dfrac{5\,dx}{12(x-2)} = \tfrac{5}{12} \int \dfrac{dx}{x-2} = \tfrac{5}{12} \log_e (x-2) + c \\[3mm] \int \dfrac{41\,dx}{24(x+4)} = \tfrac{41}{24} \int \dfrac{dx}{x+4} = \tfrac{41}{24} \log_e (x+4) + c \end{array} \right\} \quad \text{(page 159)}$$

Hence we may summarize the whole procedure thus:

$$\int \frac{(3x^2 - 7)\,dx}{x(x-2)(x+4)} = \int \left\{ \frac{7}{8x} + \frac{5}{12(x-2)} + \frac{41}{24(x+4)} \right\} dx$$

$$= \int \frac{7\,dx}{8x} + \int \frac{5\,dx}{12(x-2)} + \int \frac{41\,dx}{24(x+4)}$$

$$= \tfrac{7}{8} \int \frac{dx}{x} + \tfrac{5}{12} \int \frac{dx}{x-2} + \tfrac{41}{24} \int \frac{dx}{x+4}$$

Thus $\displaystyle\int \frac{(3x^2 - 7)\,dx}{x(x-2)(x+4)} = \tfrac{7}{8}\log_e x + \tfrac{5}{12}\log_e (x-2) + \tfrac{41}{24}\log_e (x+4) + c$

where c is the sum of the constants obtained for the three individual integrations.

(*b*) The denominator is a quadratic function of x, and if it can be factorized into the form $(x+a)(x+b)$, where a and b are constants, then the method of partial fractions can be used to integrate the given expression. To factorize $x^2 + 5x + 6$, we require two numbers, a and b, such that $a + b = 5$ and $ab = 6$. The numbers 2 and 3 satisfy these conditions so $x^2 + 5x + 6 = (x+2)(x+3)$. Thus the working of the problem in outline is

$$\int \frac{x\,dx}{x^2 + 5x + 6} = \int \frac{x\,dx}{(x+2)(x+3)} \quad \text{(by factorization)}$$

$$= \int \left\{ \frac{3}{x+3} - \frac{2}{x+2} \right\} dx \quad \text{(by partial fractions)}$$

$$= 3 \int \frac{dx}{x+3} - 2 \int \frac{dx}{x+2}$$

Thus $\displaystyle\int \frac{x\,dx}{x^2 + 5x + 6} = 3 \log_e (x+3) - 2 \log_e (x+2) + c$

So far, we have considered examples in which the denominator of the fraction to be integrated contains only linear functions of x (i.e. of the form $a + bx$), or functions of x that can be put into linear form, as in *example 5.3(b)* above. If this is not the case, the procedure must be very slightly modified.

Example 7.4

$$\text{Find} \quad \int \frac{2x\,dx}{(x+3)^2}$$

Consider first how to split $2x/(x+3)^2$ into partial fractions. At first sight, it might seem appropriate to let

$$\frac{2x}{(x+3)^2} \equiv \frac{A}{x+3} + \frac{B}{x+3}$$

However, this would be incorrect, since when the next step of multiplying both sides by $(x+3)^2$ is done, we have

$$2x \equiv A(x+3) + B(x+3) \tag{7.13}$$

Now it can be seen that *7.13* is not an identity, true for *all* values of x, since if we put $x = -3$ we have the anomaly that $-6 = 0$. If, however, we let

$$\frac{2x}{(x+3)^2} \equiv \frac{A}{x+3} + \frac{B}{(x+3)^2}$$

then, on multiplying both sides by $(x+3)^2$, we have

$$2x \equiv A(x+3) + B \tag{7.14}$$

and *7.14* is an identity. Now put $x = -3$; then $2(-3) = B$, and so $B = -6$. Using this value of B back in *7.14*, we have $2x = A(x+3) - 6$, i.e. $A(x+3) = 2x + 6$ or $A(x+3) = 2(x+3)$, and so $A = 2$. Thus

$$\frac{2x}{(x+3)^2} = \frac{2}{x+3} - \frac{6}{(x+3)^2}$$

(verify this by working from right to left). Summarizing the whole problem:

$$\int \frac{2x\,dx}{(x+3)^2} = \int \left\{ \frac{2}{x+3} - \frac{6}{(x+3)^2} \right\} dx$$

$$= 2 \int \frac{dx}{x+3} - 6 \int \frac{dx}{(x+3)^2}$$

and so

$$\int \frac{2x\,dx}{(x+3)^2} = 2 \log_e (x+3) + \frac{6}{x+3} + c$$

See page 107 for the integration of the second term.

The area under a curve and the definite integral

You may be wondering why, if integration is the converse of differentiation, the process is not given a name like 'de-differentiation'. The answer to this lies in the remark made at the beginning of the chapter, that the process of integration can be applied in two different situations. We have already examined one of these, the concept of integration as the reverse of differentiation. The other concept is implied in the name of the process, 'integration', which means 'adding together', and we now turn to examine this aspect of integration.

We begin by stating and proving a basic fact. In Fig. 7.1 the area enclosed by the curve of $y = f(x)$, the x-axis, and the vertical lines at $x = a$ and $x = b$ (i.e. the area enclosed by the curve AB and the lines BC, CD, and DA) can be

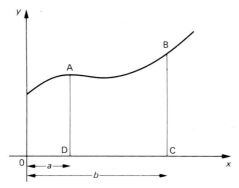

Fig. 7.1   The area under a curve (see page 112).

represented by an integral with $f(x)$ as integrand; and moreover, the area may be evaluated precisely.

To prove these statements, look first at Fig. 7.2. It is required to find the area enclosed by the curve of $y = f(x)$, the x-axis, and the vertical lines at $x = a$ and $x = b$. Imagine now that the area is dissected into a very large number of very thin vertical strips, each of width δx. In Fig. 7.2 one of these strips is shown (highly magnified in width) whose left-hand edge PM is distant x from the origin. The height of this edge is y, i.e. $f(x)$, since PM is distant x from the origin. The area of the rectangle PMNR is $y\delta x$, and, since the area enclosed by PQR is very small, the area of the complete strip PMNQ is approximately equal to $y\delta x$.

Now let us define some quantities in Fig. 7.2 and give them symbols. Let the area required, ABCD, be S and let the area APMD be s. Further, since δx is

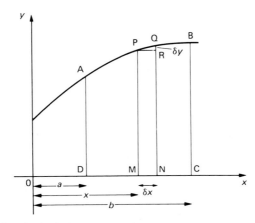

Fig. 7.2   A typical vertical strip in an area under a curve.

very small, the small element of area given by the strip is, as we have seen, $\simeq y\delta x$. Since area PQMN can be regarded as a small increase to area APMD owing to a small increase δx in x, we may call the area PQMN δs.

$$\text{Hence} \qquad \delta s \simeq y\delta x \qquad\qquad (7.15)$$

The narrower the strips, the smaller δx becomes, and the approximation *7.15* becomes more accurate. So we may say that,

$$\text{as} \qquad \delta x \to 0, \qquad \text{so} \qquad \delta s \to y\delta x$$

$$\text{or} \qquad \frac{\delta s}{\delta x} \to y$$

Writing this in proper limit notation, we have

$$\lim_{\delta x \to 0} \left(\frac{\delta s}{\delta x} \right) = y \qquad\qquad (7.16)$$

But in keeping with previous notation, *equation 5.5*, where we wrote

$$\lim_{\delta x \to 0} \left(\frac{\delta y}{\delta x} \right) = \frac{dy}{dx}$$

we can write in the present instance

$$\lim_{\delta x \to 0} \left(\frac{\delta s}{\delta x} \right) = \frac{ds}{dx} \qquad\qquad (7.17)$$

Since the left-hand sides of *7.16* and *7.17* are equal to one another, we may equate the right-hand sides of these two expressions:

$$\frac{ds}{dx} = y$$

$$\text{Cross-multiplying gives} \qquad ds = y\,dx$$

$$\text{and so } s, \text{ the area APMD, is given by} \qquad s = \int y\,dx$$

$$\text{or, since } y = f(x), \qquad s = \int f(x)\,dx$$

Now assume that $\qquad \int f(x)\,dx = F(x) + c$

where $d\{F(x)\}/dx = f(x)$ (page 103), and c is a constant of integration:

$$\text{therefore} \qquad s = F(x) + c \qquad\qquad (7.18)$$

We may regard the measurement of area s as beginning from the line AD. Hence at line AD, where $x = a$, there is no area, and so $s = 0$. So substituting for s and x in *7.18*, we have

$$0 = F(a) + c$$

$$\text{which gives} \qquad c = -F(a)$$

Substituting back into *7.18* for c gives

$$s = F(x) - F(a) \qquad\qquad (7.19)$$

Now consider the whole required area ABCD ($=S$); for this area $x = b$ in *7.19*. On substituting for x in *7.19*, we have

$$S = F(b) - F(a) \qquad\qquad (7.20)$$

Equation 7.20 is the required result, and can be stated as follows: 'the area under the curve of $y = f(x)$, between the limits $x = a$ and $x = b$, is the value of $F(x)$ when $x = b$ minus the value of $F(x)$ when $x = a$; $F(x)$ is the integral of $f(x)$'. Notice that *7.20* contains no constant of integration; S is the difference between $F(b)$, which does contain such a constant, and $F(a)$ which contains the same constant, and thus cancels out.

Evaluation of a definite integral

In working problems on areas under curves, the following notation is generally used. The right-hand side of *7.20* is written as

$$\left[F(x) \right]_a^b \qquad\qquad (7.21)$$

that is, the integrated function is put in square brackets and the upper and lower limits of x at which F is to be evaluated are written as shown. The integral of the function of the original curve which gives rise to *7.21* is written as

$$\int_a^b f(x)\,dx \qquad\qquad (7.22)$$

which is the integral of $f(x)\,dx$ as written in the previous chapter, but with the upper and lower limits shown. Therefore, as *7.20*, *7.21*, and *7.22* are merely different ways of expressing the same thing, we have, in the order they arise in working through a problem

$$S = \int_a^b f(x)\,dx = \left[F(x)\right]_a^b = F(b) - F(a) \tag{7.23}$$

$$\text{where} \qquad d\{F(x)\},\,dx = f(x)$$

The symbol $\int_a^b f(x)\,dx$ is clearly somewhat different from the symbol $\int f(x)\,dx$, for

$$\int f(x)\,dx = F(x) + c \tag{7.24}$$

$$\text{and} \qquad \int_a^b f(x)\,dx = F(b) - F(a) \tag{7.25}$$

However, in both instances $d\{F(x)\}/dx = f(x)$ and so integration is actually carried out in the same way in each case. But in *7.25* there is the further step of inserting limits into the integrated function. Since *7.25* does not involve the unknown constant of integration, $\int_a^b f(x)\,dx$ is known as the **definite integral**; $\int f(x)\,dx$, you will recall, is the indefinite integral.

Example 7.5
Find the areas under the curves indicated:

(a) of the function $y = x^2$ between the limits of 1 and 3;
(b) of the function $y = \sqrt{(2x + 1)}$ between limits of 0 and 4.

(a) The situation is shown in Fig. 7.3, and the value of the shaded area is required, S. Since the curve represents the function $y = x^2$, and the vertical limits of the area are $x = 1$, and $x = 3$, the required area is given by

$$S = \int_1^3 x^2\,dx$$

$$= \left[\frac{x^3}{3}\right]_1^3$$

$$= \frac{(3)^3}{3} - \frac{(1)^3}{3} = 9 - \tfrac{1}{3} = 8\tfrac{2}{3}$$

Hence the shaded area is $8\tfrac{2}{3}$ square units.

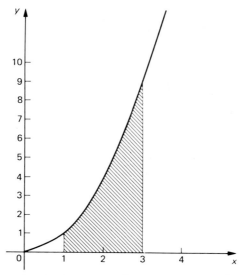

Fig. 7.3   The area under the curve of $y = x^2$ between the limits of $x = 1$ and $x = 3$ (see *example 7.4a*).

(*b*) The required area is given by

$$S = \int_0^4 \sqrt{(2x + 1)}\, dx = \int_0^4 (2x + 1)^{1/2}\, dx = \left[\frac{(2x + 1)^{3/2}}{3} \right]_0^4$$

(See *equation 7.9* for the integration of this function.)

Hence $S = \dfrac{\sqrt{\{2(4) + 1\}^3}}{3} - \dfrac{\sqrt{\{2(0) + 1\}^3}}{3} = 9 - \tfrac{1}{3} = 8\tfrac{2}{3}$ square units

Another difference should now be evident between a definite and an indefinite integral: while the latter merely gives rise to another function, the former gives a numerical value.

For example $\int x^2\, dx = \tfrac{1}{3}x^3 + c$

whereas $\int_1^3 x^2\, dx = 8\tfrac{2}{3}$

A definite integral can also exist, or be thought of, on its own without any reference to an area under a curve, just as a derivative does not have to be associated with the gradient of a curve, or a function of x does not have to be identified with a curve in the x–y plane. One can merely say, 'Evaluate $\int_1^3 x^2 \, dx$', and the answer is $8\frac{2}{3}$,

It cannot be too strongly emphasized that the foregoing material of this chapter is very important as background theory to many leading aspects of quantitative biology. The discussion, notations involved, and the proof that the area under a curve can be evaluated by a definite integral, should be thoroughly assimilated and understood before proceeding.

'Negative' areas

Consider the areas shown in Fig. 7.4, each of which will be given by $\int_a^b f(x) \, dx$. In Fig. 7.4(a) and (b), the area is on the upper side of the x-axis

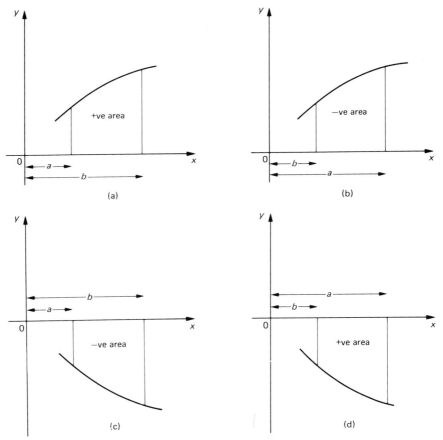

Fig. 7.4 'Positive' and 'negative' areas under curves.

(positive y-values); further, in (a) the upper limit, b, is greater than the lower limit, a, i.e. $b > a$. This situation is identical with our previous examples, and as we have seen, the definite integral yields a numerical value which is positive. In (b) is shown the same area, but now the upper limit, b, is less than the lower limit, a, i.e. $b < a$. Under these conditions, the definite integral will yield the same *numerical* result as in (a) but opposite in sign, i.e. negative. Nevertheless, it is obvious that the same area is being measured, as is shown by the result being numerically the same; but since the limits are reversed, the sign of the definite integral is likewise reversed. The terms 'upper' and 'lower', when applied to limits of a definite integral, refer to their positions above and below the integral sign and have nothing to do with their values relative to one another.

Now look at Fig. 7.4(c) and (d), where the area lies in the negative range of y. If $b > a$, as in Fig. 7.4(c), the definite integral is negative, and if $b < a$, as in Fig. 7.4(d), the result is positive. These facts demonstrate further the difference between the purely mathematical concept of the definite integral, and its physical representation as the area under a curve. The latter must always be a positive quantity, whereas the former may be positive or negative. But an important practical consideration arises if the curve of $y = f(x)$ crosses the x-axis between the limits of integration. The total numerical area must then be evaluated as the sum of two separate areas – that above the x-axis and that below the x-axis – each area being assumed positive.

Area between two curves

Suppose it is required to find the shaded area in Fig. 7.5, i.e. the area between the curve of $y = f(x)$ and the curve of $y = \phi(x)$ between the limits of $x = a$ and $x = b$. Now the area under the curve of $y = f(x)$ is given by $\int_a^b f(x)\,dx$, and the area under the curve of $y = \phi(x)$ is given by $\int_a^b \phi(x)\,dx$. Clearly, the shaded

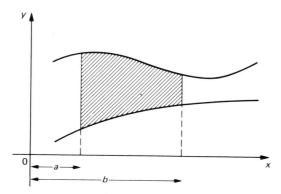

Fig. 7.5   The area between two curves, between the limits of $x = a$ and $x = b$. The upper curve is that of $y = f(x)$, and the lower curve depicts the function $y = \phi(x)$.

area is equal to the area under the curve of $y = f(x)$ minus the area under the curve of $y = \phi(x)$,

i.e. $$S = \int_a^b f(x)\,dx - \int_a^b \phi(x)\,dx \qquad (7.26)$$

Example 7.6
(a) Find the area enclosed by the curve $y = 2x - x^2$ and the straight line $y = 2$ between the limits of $x = 0$ and $x = 2$.
(b) Find the area enclosed by the curves $y = 4x - x^2$ and $y = x^2 - 2x + 2.5$.

(a) A diagram should always be drawn first, and the present situation is shown in Fig. 7.6(a), with the area to be evaluated shaded. Using 7.26, the area is given by

$$S = \int_0^2 2\,dx - \int_0^2 (2x - x^2)\,dx$$

$$= \left[2x - (x^2 - x^3/3) \right]_0^2$$

$$= [2(2) - \{(2)^2 - (2)^3/3\}] - [2(0) - \{(0)^2 - (0)^3/3\}]$$

$$= 4 - 4 + \tfrac{8}{3} = 2\tfrac{2}{3}$$

(b) The situation is shown in Fig. 7.6(b), and since no limits are specified, it is evident that the whole enclosed area between the points of intersection of the

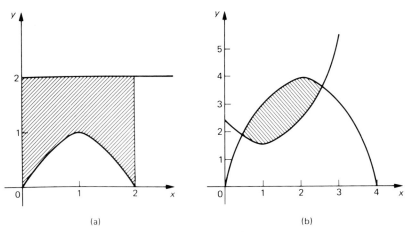

(a)   (b)

Fig. 7.6   (a) The area between the curve of the function $y = 2x - x^2$ and the straight line $y = 2$, between the limits of $x = 0$ and $x = 2$ (see *example 8.2*a). (b) The area enclosed by the curves $y = 4x - x^2$ and $y = x^2 - 2x + 2.5$ (see *example 8.2*b).

two curves is required, and so these must be found first. The points of intersection lie where both equations are satisfied together, and so are found by equating the right-hand sides of the equations of the two curves, giving

$$x^2 - 2x + 2.5 = 4x - x^2$$

$$\text{i.e.} \qquad 2x^2 - 6x + 2.5 = 0 \qquad\qquad (7.27)$$

Solving *7.27* for x will yield the abscissae of the points of intersection, which is all we need to set limits on the integral. Using the formula for solving a quadratic equation (see *equations 4.1* and *4.2*), we have

$$x = \frac{6 \pm \sqrt{(36 - 20)}}{4} = \frac{6 \pm \sqrt{16}}{4} = \frac{6 \pm 4}{4}$$

So either $\quad x = \tfrac{1}{2} \quad$ or $\quad x = 2\tfrac{1}{2}$

Hence the two curves intersect at the points whose x co-ordinates are $\tfrac{1}{2}$ and $2\tfrac{1}{2}$, and so these two values of x are the lower and upper limits of integration respectively. Therefore, the required area is given by

$$S = \int_{1/2}^{5/2} (4x - x^2)\,dx - \int_{1/2}^{5/2} (x^2 - 2x + 2.5)\,dx$$

$$= \left[2x^2 - x^3/3 - x^3/3 + x^2 - 2.5x \right]_{1/2}^{5/2}$$

$$= \left[3x^2 - 2x^3/3 - 5x/2 \right]_{1/2}^{5/2}$$

$$= \left\{ 3\left(\frac{5}{2}\right)^2 - 2\left(\frac{5}{2}\right)^3/3 - 5\left(\frac{5}{2}\right)/2 \right\} - \left\{ 3\left(\frac{1}{2}\right)^2 - 2\left(\frac{1}{2}\right)^3/3 - 5\left(\frac{1}{2}\right)/2 \right\}$$

$$= \frac{3(25)}{4} - \frac{2(125)}{3(8)} - \frac{24}{4} - \frac{3}{4} + \frac{2}{24} + \frac{5}{4} = \frac{482 - 418}{24} = \frac{64}{24} = \frac{8}{3} = 2\tfrac{2}{3}$$

Hence the area enclosed by the curves $y = 4x - x^2$ and $y = x^2 - 2x + 2.5$ is $2\tfrac{2}{3}$ square units.

Integration as a summation

Consider the situation shown in Fig. 7.2, which illustrated the proof of the area under a curve. The proof involved dividing the area up into a large number of vertical strips and considering any one of these. A typical strip is shown in Fig. 7.2, PQNM, and its area is approximately $y\delta x$. Now the whole area

ABCD is made up of the sum of the area of all the vertical strips between $x = a$ and $x = b$. Thus

$$S \simeq \sum_{x=a}^{b} y \, \delta x = \sum_{x=a}^{b} f(x) \, \delta x$$

(see page 139 for details concerning the Σ-notation); or, since the area of each strip approaches $y\delta x$ more accurately as $\delta x \to 0$,

$$S = \lim_{\delta x \to 0} \left\{ \sum_{x=a}^{b} f(x) \, \delta x \right\}$$

But we have already shown that $S = \int_a^b f(x) \, dx$, so

$$\int_a^b f(x) \, dx = \lim_{\delta x \to 0} \left\{ \sum_{x=a}^{b} f(x) \, \delta x \right\} \qquad (7.28)$$

Relationship *7.28* is very important, as it shows that if we have a situation wherein a very large number of very small elements need to be added together, the summation may be replaced by an integral.

Example 7.7

Find the area of a circle of radius r.

Let the circle be dissected into a large number of very thin rings, each of width δx, as shown in Fig. 7.7. Consider a typical ring whose inner edge is distant x from the centre of the circle. The circumference of the inner edge of

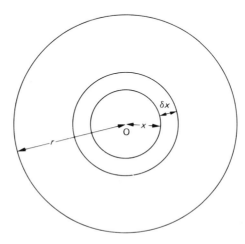

Fig. 7.7   A circle of radius r dissected into thin rings of width δx.

the ring is $2\pi x$. Since δx is small, the area of the ring is approximately $2\pi x \delta x$, as can be appreciated by imagining the ring to be cut out and straightened, when it would approximate to a rectangle of length $2\pi x$ and width δx. Hence the area of the circle, S, is equal to the sum of all such rings, i.e.

$$S = \lim_{\delta x \to 0} \left(\sum_{x=0}^{r} 2\pi x \, \delta x \right) = \int_{0}^{r} 2\pi x \, \delta x$$

$$= 2\pi \int_{0}^{r} x \, dx = 2\pi \left[x^2/2 \right]_{0}^{r}$$

$$= \pi \left[x^2 \right]_{0}^{r} = \pi r^2$$

The mean value of a function

The mean, or average, value of a set of n quantities: $y_1, y_2, y_3, \ldots, y_n$, is given by

$$\bar{y} = \frac{y_1 + y_2 + y_3 + \cdots + y_n}{n} = \frac{\Sigma y}{n}$$

Can we find the mean value of a function (of x) between two specified values of x? Look at Fig. 7.8 which shows a graph of the function of x, i.e. of $y = f(x)$, and suppose it is desired to find the mean value of $f(x)$ between the values of $x = a$ and $x = b$. Imagine the interval to be divided up into sub-intervals as shown, with each sub-interval being of width δx. The more sub-intervals there

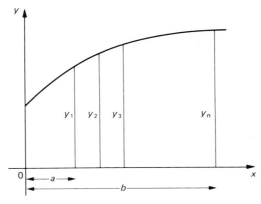

Fig. 7.8   Some ordinates between those at $x = a$ and $x = b$.

are, the smaller is δx, and as $n \to \infty$, so $\delta x \to 0$. Now the mean value of the curve is given by

$$\bar{y} = \lim_{n \to \infty} \left(\frac{y_1 + y_2 + \cdots + y_n}{n} \right) = \lim_{n \to \infty} \left\{ \frac{(y_1 + y_2 + \cdots + y_n)\,\delta x}{n\,\delta x} \right\}$$

the last term above being obtained from the previous one by multiplying both numerator and denominator by δx.

$$\text{Hence} \qquad \bar{y} = \lim_{n \to \infty} \left(\frac{\sum\limits_{x=a}^{b} y\,\delta x}{n\,\delta x} \right)$$

Now since $y = f(x)$, $(n-1)\,\delta x = b - a$, and $\delta x \to 0$ as $n \to \infty$, the above expression becomes

$$\bar{y} = \lim_{\delta x \to 0} \left\{ \frac{\sum\limits_{x=a}^{b} f(x)\,\delta x}{b - a + \delta x} \right\} = \frac{\lim\limits_{\delta x \to 0} \left\{ \sum\limits_{x=a}^{b} f(x)\,\delta x \right\}}{\lim\limits_{\delta x \to 0} \{ b - a + \delta x \}}$$

The last term follows because the limit of a quotient equals the quotient of two limits. Therefore

$$\bar{y} = \frac{\int_a^b f(x)\,\mathrm{d}x}{b - a} = \frac{\text{area under the curve}}{\text{interval of } x} \qquad \text{(page 122)}$$

which is usually written as

$$\bar{y} = \frac{1}{b - a} \int_a^b f(x)\,\mathrm{d}x \qquad\qquad (7.29)$$

Example 7.8
Find the mean value of the function x^2 between the limits of $x = 1$ and $x = 3$. We shall denote the mean value of a function $f(x)$ by $\overline{f(x)}$. Then

$$\overline{f(x)} = \frac{1}{3 - 1} \int_1^3 x^2\,\mathrm{d}x - \frac{1}{2} \left[\frac{x^3}{3} \right]_1^3 = 4\tfrac{1}{3}$$

Note that $f(1) = 1$ and $f(3) = 9$, but the mean value of $f(x)$ between limits $x = 1$ and $x = 3$ is $4\tfrac{1}{3}$. As with a mean value in statistics, the calculation has given due weight to the distribution of values of the function within the interval 1 to 3 (see Fig. 7.3).

Numerical integration

In the previous chapter it was stated that it was not possible to integrate certain functions. Now this does not mean to say that for such a function, $f(x)$, its integral, $\int f(x)\,dx$, does not exist. What it does mean is that for integration, which is defined as usual as

$$\int f(x)\,dx = F(x) + c$$

no known method will enable us to find $F(x)$.

However, if we put $y = f(x)$, then for most functions a graph could be drawn and an area under the curve of $y = f(x)$ between two limits $x = a$ and $x = b$ defined. Hence we know that the definite integral, $\int_a^b f(x)\,dx$, exists because the area just defined is the physical representation of the integral; but one is unable to evaluate the definite integral by direct analytical means.

As an example, consider the function 2^{-x^2}; this cannot be integrated by any known method. However, the curve of $y = 2^{-x^2}$ is shown in Fig. 7.9, and an area under the curve is defined between $x = 1$ and $x = 2$. This area is identified with the definite integral $\int_1^2 2^{-x^2}\,dx$, but the latter cannot be used for evaluating the area. If, however, some other method could be used for at least *estimating* the shaded area, the result would also be an approximation of the value of the

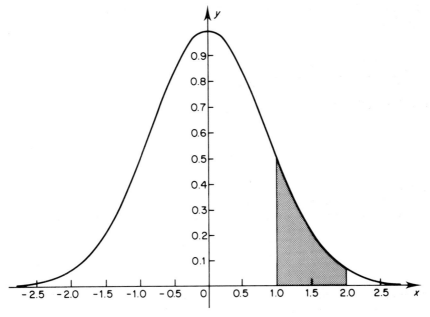

Fig. 7.9 Curve of the function $y = 2^{-x^2}$.

definite integral. Such procedures do exist, and are called methods of **numerical integration**: two will be described here.

Counting squares

This method is very elementary, and laborious. It consists of drawing a very accurate graph of the function over the range of x concerned, drawing the vertical lines at the limits, and then counting the squares of the graph paper within the required area. Assuming that the graph is drawn on metric paper, every 1 mm square would need to be counted. This would present no problems on the three straight sides of the area (apart from tedium!), but much subjectivity is required along the edge defined by the curve itself.

Simpson's Rule

This is the most well-known method of numerical integration; it is capable of giving good results, and is not too laborious to carry out. The rule will be stated below, but not proved.

Suppose we require to evaluate $\int_a^b f(x)\,dx$ by Simpson's Rule. First, consider a graph of $y = f(x)$ with the area marked out between the limits $x = a$ and $x = b$ (Fig. 7.10). Then divide the area into an *even* number of strips of width Δx (see page 68 for discussion of the notations Δx and δx). There is no question here of the strips being very narrow and very large in number; in Fig. 7.10 there are only 4 strips. Since a, b, and Δx are known, it follows that the x-values (abscissae) of the vertical lines defining the strips can be evaluated, and by substituting these abscissae into $f(x)$, the corresponding y-values (ordinates) can be calculated. Simpson's Rule then gives the area as

$$S \simeq \text{one-third the width of a strip} \begin{pmatrix} \text{first ordinate + last ordinate +} \\ \text{twice the sum of the remaining odd} \\ \text{ordinates + four times the sum of} \\ \text{the even ordinates} \end{pmatrix}$$

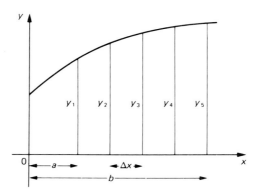

Fig. 7.10   Simpson's Rule to find the area under the curve.

For the situation shown in 7.10, where there are four strips, the formula for Simpson's Rule is

$$S \simeq \frac{\Delta x}{3} \{y_1 + y_5 + 2y_3 + 4(y_2 + y_4)\} \qquad (7.30)$$

Example 7.9

Evaluate $\int_1^2 2^{-x^2}$

by Simpson's Rule, using (a) 4 strips, (b) 10 strips.

(a) It is useful to calculate the ordinates by means of a table.

x	$-x^2$	$2^{-x^2} = y$	$2y$ (where appropriate)	$4y$ (where appropriate)
1.00	−1.0000	0.5000	—	—
1.25	−1.5625	0.3386	—	1.3544
1.50	−2.2500	0.2102	0.4204	—
1.75	−3.0625	0.1197	—	0.4788
2.00	−4.0000	0.0625	—	—

Thus

$$S \simeq \frac{0.25}{3} \{(0.5000 + 0.0625) + 0.4204 + (1.3544 + 0.4788)\}$$

$$\simeq \frac{0.25 \times 2.8161}{3} = 0.234675$$

(b)

x	$-x^2$	$2^{-x^2} = y$	$2y$ (where appropriate)	$4y$ (where appropriate)
1.0	−1.00	0.5000	—	—
1.1	−1.21	0.4323	—	1.7292
1.2	−1.44	0.3686	0.7372	—
1.3	−1.69	0.3099	—	1.2396
1.4	−1.96	0.2570	0.5140	—
1.5	−2.25	0.2102	—	0.8408
1.6	−2.56	0.1696	0.3392	—
1.7	−2.89	0.1349	—	0.5396
1.8	−3.24	0.1058	0.2116	—
1.9	−3.61	0.0819	—	0.3276
2.0	−4.00	0.0625	—	—

Thus

$$S \simeq \frac{0.1}{3} \{(0.5000 + 0.0625) + (0.7372 + 0.5140 + 0.3392 + 0.2116)$$

$$+ (1.7292 + 1.2396 + 0.8408 + 0.5396 + 0.3276)\}$$

$$\simeq \frac{0.1 \times 7.0413}{3} = 0.23471$$

Hence $\int_1^2 2^{-x^2} \, dx = 0.2347$ (correct to 4 decimal places).

EXERCISES

1. Find the following indefinite integrals:

(a) $\int 6x^2 \, dx$ (b) $\int (5/x^2) \, dx$ (c) $\int (x^3 - a^3) \, dx$ (d) $\int (x - 1/x)^2 \, dx$

(e) $\int (1 + x^2) \sqrt{x} \, dx$ (f) $\int (2x - 1)^5 \, dx$ (g) $\int 1/\sqrt[3]{(1 - 4x)} \, dx$

2. Integrate the following, using the method of partial fractions

(a) $\int \dfrac{(3x + 4) \, dx}{(x - 2)(x + 3)}$ (b) $\int \dfrac{(x^2 - 3x + 3) \, dx}{(x - 1)(x - 2)(x - 3)}$

3. Evaluate (a) $\int_1^3 (2x + x^2) \, dx$ (b) $\int_4^9 \{(x + 1)/\sqrt{x}\} \, dx$

4. Evaluate $\int_1^2 dx/x^2$ by the following methods.
(a) Draw an accurate graph of the functions $y = 1/x^2$ between $x = 1$ and $x = 2$, using as many intermediate x-values and as large a scale as practicable. Count the squares of the graph paper lying within the area enclosed by the curve, the x-axis, and the vertical lines $x = 1$ and $x = 2$. Express the results in square units.
(b) Use Simpson's Rule with 4 strips.
(c) Use Simpson's Rule with 10 strips.
(d) Integrate directly.
Compare the four results.

5. Find the mean value of the following functions between the limits stated:

(a) $2x + x^2$ between $x = 1$ and 3,

(b) $(x + 1)/\sqrt{x}$ between $x = 4$ and 9.

8
Differential equations and mathematical series

The two topics to be introduced in this chapter bear no relation to one another; but a brief look at differential equations at this point will round off our systematic study of the calculus, and the idea of a mathematical series is conveniently placed here as one such series is required for the next chapter.

Steps in solving a differential equation

The idea of a differential equation was introduced in Chapter 6, and pages 88 to 95 should be re-read at this point before proceeding.

In the discussion of the growth of a microbial culture, the aim was to obtain a mathematical function describing the increase of culture size with time, i.e. $n = f(t)$, by setting up a mathematical model of growth based on supposed features of microbial growth. The model considered two broad features of the dynamics of culture growth, and finally described the way that the rate of change of number of organisms was related to the number of organisms present in the culture at a particular time. The relationship between rate of growth and organism number was a differential equation, *6.12*, which was

$$\frac{dn}{dt} = kn \left(1 - \frac{n}{a} \right)$$

or, in a form more convenient for our present purpose,

$$\frac{dn}{dt} = \frac{kn}{a} (a - n) \qquad (8.1)$$

This is a function of the form $dn/dt = f(n)$, but the function we require, $n = f(t)$, can be obtained by 'solving' the differential equation 8.1. To solve a differential equation, we evidently need to eliminate the derivative which, of course, is done by integration.

The subject of differential equations is a difficult one, and only a few introductory topics are introduced here. Unfortunately, the more interesting equations of biological relevance are beyond the scope of this elementary introduction; but useful differential equations from a biological viewpoint presented in *A Biologist's Advanced Mathematics*, in a chapter concerned with a more systematic study of differential equations. First, *equation 8.1* will be solved, since this will complete the model of the microbial culture.

Although the core feature in the solution of a differential equation is integration, there is usually some preparatory work to be done first, and some 'tidying up' operations afterwards: so a number of distinct stages can usually be recognized. For *8.1*, these stages are as follows.

(*i*) Re-arrange, to obtain terms in n and dn on the left-hand side, and dt on the right:

$$\frac{dn}{n(a-n)} = \frac{k\,dt}{a}$$

(*ii*) Now, each side is integrated; the left-hand side with respect to n, and the right-hand side with respect to t. However, the left-hand side cannot be integrated as it stands, but must be split into partial fractions first (pages 107 to 112).

$$\text{Hence} \qquad \frac{dn}{an} + \frac{dn}{a(a-n)} = \frac{k\,dt}{a}$$

and the a's cancel, giving

$$\frac{dn}{n} + \frac{dn}{a-n} = k\,dt \qquad (8.2)$$

If this is your first reading of the book, and you have missed out the section on 'integration by partial fractions' in Chapter 7, just accept the above result for the time being. It is, however, very easy to see that $1/n + 1/(a-n) = a/n(a-n)$: this is all we have done to the left-hand side of the equation. Now integrate *8.2* (see pages 157 and 159):

$$\log_e n - \log_e (a-n) = kt + c$$

where c is the sum of the constants of integration of each of the separate integrals (separate terms) of *8.2*.
(*iii*) Now it merely remains to write the answer in the most useful form. First, we have

$$\log_e\left(\frac{n}{a-n}\right) = kt + c$$

Next, take exponentials (anti-natural logarithms) of both sides:

$$\frac{n}{a-n} = e^{(kt+c)}$$

Now invert both sides, and split the right-hand side using the First Law of indices:

$$\frac{a-n}{n} = \frac{1}{e^{kt}\,e^{c}}$$

i.e. $$\frac{a}{n} - 1 = e^{-kt}\,e^{-c}$$

Put $b = e^{-c}$, so that $$\frac{a}{n} = 1 + b\,e^{-kt}$$

Invert both sides again: $$\frac{n}{a} = \frac{1}{1 + b\,e^{-kt}}$$

and so $$n = \frac{a}{1 + b\,e^{-kt}} \qquad (8.3)$$

which is the logistic function (page 162). It is of course quite suitable for describing, at least approximately, the growth of a culture of micro-organisms.

Now that we have seen an example of the occurrence, use, and solution of a differential equation, we shall examine these equations in greater detail.

Definitions

Order

A differential equation may contain derivatives of any order, dy/dx, $d^2 y/dx^2$, $d^3 y/dx^3$, etc., and the order of a differential equation is the highest derivative contained in the equation. Hence

$$a\frac{d^2 y}{dx^2} + b\frac{dy}{dx} + cy = 0$$

is a second-order differential equation.

Degree

The degree of a differential equation is the power of the highest derivative; thus

$$\left(\frac{d^2 y}{dx^2}\right)^3 - a\frac{dy}{dx} = c$$

is a second-order equation of the third degree, while

$$a\left(\frac{d^2 y}{dx^2}\right)^2 + b\left(\frac{dy}{dx}\right)^3 + cy - d = 0$$

is a second-order equation of the second degree, because the power of the highest derivative is 2.

General solution

When the solution of a differential equation contains a constant of integration, of unknown value, such as b $(=e^{-c})$ in *equation 8.3*, the solution is known as a general solution.

Particular solution

Sometimes, when additional information is available, the constant of integration can be evaluated. Then, the solution of the differential equation is a particular solution. For instance, let the initial inoculum in our microbial culture be n_0. Thus when $t = 0$, $n = n_0$. With this information, we can substitute in *equation 8.3* for the unknown constant of integration:

$$n_0 = \frac{a}{1 + b}$$

i.e. $\quad b = a/n_0 - 1$

Thus *8.3* becomes $\qquad n = \dfrac{a}{1 + \left(\dfrac{a}{n_0} - 1\right) e^{-kt}}$ $\qquad$ (8.4)

and so *8.4* is a particular solution of the differential equation *8.1*.

Figure 8.1 shows the curves of three different logistic functions of the form of *8.3*, in which the constants a and k are identical from one function (curve) to another. However, the constant b differs for each, and since $b = a/n_0 - 1$ this

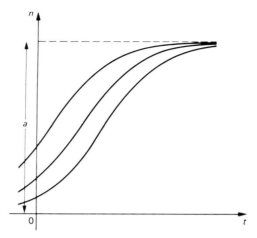

Fig. 8.1   Curves of the logistic function $n = a/(1 + b\,e^{-kt})$ differing only in the value of b.

means that n_0 also differs. These functions could describe the growth of three different cultures of micro-organisms of the same species and under identical conditions; only the initial sizes of the inoculum, n_0, differ from one to another. On the other hand, the differential equation *8.1* describing the underlying dynamic situation which gives rise to the curves shown in Fig. 8.1 is the same for all three of the curves, since it contains the constants a and k only. This confirms that, under this model, the *dynamics* of culture growth are not affected by the size of the inoculum.

The general solution of *equation 8.1* can be represented by an infinite number of curves of the form shown in Fig. 8.1, and each has the property that at any point the gradient is given by

$$\frac{dn}{dt} = kn\left(1 - \frac{n}{a}\right)$$

the differential equation with which we started. Any particular curve, such as one of those shown in Fig. 8.1 with a specified value of n_0 (and hence of b, and so in turn c, the constant of integration), represents a particular solution of the differential equation *8.1*.

Equations where the variables are separable

Equations of the form $dy/dx = f(x)$

These are very straightforward to solve, provided that $f(x)$ is integrable, since

$$y = \int f(x)\,dx$$

Example 8.1

$$\text{Solve} \qquad \frac{dy}{dx} = bx$$

$$\text{We have} \qquad y = b \int x \, dx$$

$$\text{Hence} \qquad y = bx^2/2 + c$$

Equations of the form $dy/dx = f(y)$

In these cases a cross-multiplication is first made to get all terms in y and dy on to the left-hand side, and dx on to the right-hand side of the equation,

$$\text{i.e.} \qquad \frac{dy}{f(y)} = dx$$

Any multiplying constant in $f(y)$ is best allowed to remain on the right-hand side. Then the left-hand side is integrated with respect to y, and the right-hand side with respect to x.

Example 8.2

$$\text{Solve} \qquad \frac{dy}{dx} = by$$

$$\text{Cross-multiply:} \qquad \frac{dy}{y} = b \, dx$$

$$\text{Integrating,} \qquad \log_e y = bx + c$$

Take exponentials (i.e. anti-natural logarithms) of both sides,

$$y = e^{bx} e^c$$

Put $a = e^c$, then we have

$$y = a e^{bx}$$

which is the exponential function.

Equations of the form $f(y)\,dy = \phi(x)\,dx$

This is a special case of the form $dy/dx = F(x, y)$, where the derivative is a function of both x and y. However, if we are to solve this equation, it must be capable of reduction to the form $f(y)\,dy = \phi(x)\,dx$, with all terms in y and dy on one side of the equation and all terms in x and dx on the other side.

Example 8.3

$$\text{Solve} \qquad \frac{dy}{dx} = \frac{(1+x)\,y}{x^2(1-y)}$$

$$\text{Cross-multiply} \qquad \frac{(1-y)\,dy}{y} = \frac{(1+x)\,dx}{x^2}$$

$$\text{i.e.} \qquad \frac{dy}{y} - dy = \frac{dx}{x^2} + \frac{dx}{x}$$

$$\text{Integrating,} \qquad \log_e y - y = -\frac{1}{x} + \log_e x + c$$

This expression cannot be simplified any further: often we cannot set the solution of a differential equation in the form $y = f(x)$ explicitly. Nevertheless, the last line in this example is still the general solution of the differential equation, since the derivative has been eliminated.

Mathematical series

A *series* is a succession of quantities which, after the first, are formed according to a common law. Each of the quantities is called a *term* of the series. For example:

  3, 8, 13, 18, ... is a series, as each term is formed by adding 5 to the preceding term;
  $a, a + d, a + 2d,$ ... is a series, since each term is formed from the previous term by adding d;
  4, 12, 36, 108, ... is a series, since each term is formed by multiplying the preceding term by 3;
  $1, 2x, 4x^2, 8x^3,$ ... is a series, each term formed by multiplying the previous term by $2x$.

Series are very important in mathematics. As biologists working through this book, our most immediate need of a series is to define e, the base of natural logarithms, which we shall do in the next chapter. But with biology coming to rely more on numerical methods than previously, the use of electronic computers is increasing in many fields of biology, and programming a computer often involves the use of series. You may already have met two kinds of series in your school-days, the arithmetic and the geometric; we shall revise the former briefly and develop some notation, and then go on to study the geometric series in greater depth.

The arithmetic series

An **arithmetic series** is a series of terms that increase by a constant amount, which may be positive or negative. The following are examples of arithmetic series:

$1, 2, 3, 4, \ldots$	common difference $= +1$
$\frac{1}{4}, \frac{3}{2}, \frac{7}{4}, 2, \ldots$	common difference $= +\frac{1}{4}$
$6, 3, 0, -3, \ldots$	common difference $= -3$
$-2, -1\frac{1}{2}, -1, -\frac{1}{2}, \ldots$	common difference $= +\frac{1}{2}$

General notation

A general notation will now be developed for the arithmetic series, and a similar development may be made for any series. Consider an arithmetic series where the first term is a and whose common difference is d. Then

1st term	$= a$
2nd term	$= a + d$
3rd term $= (a + d) + d$	$= a + 2d$
4th term $= (a + 2d) + d$	$= a + 3d$
$\vdots$	$\vdots$
ith term (general term)	$= a + (i - 1)\,d$
$\vdots$	$\vdots$
nth term (last term)	$= a + (n - 1)\,d$

The ith term is known as the **general term** because it indicates the general structure of any term in the series; any particular term can be evaluated from the given form of the ith term. For example, the 4th term of an arithmetic series whose first term, a, is 6, and whose common difference, d, is -3, is given by $6 + (4 - 1)(-3) = 6 + (3)(-3) = 6 - 9 = -3$. The nth term is by convention the **last term** in the series.

Sum of n terms

The most important feature of a series is the sum of a number of its terms, and it is very convenient to be able to find a formula for this sum. This is easy to do in the case of the arithmetic series.

Theorem 8.1   The sum S_n, of n terms of an arithmetic series, whose first term is a and whose common difference is d, is given by

$$S_n = \frac{n}{2} \{2a + (n-1)\, d\}$$

The last term of the series is $a + (n-1)\, d$; hence

$$S_n = a + (a+d) + (a+2d) + \cdots + a + (n-1)\, d \qquad (8.5)$$

Reversing the series, we have

$$S_n = \{a + (n-1)\, d\} + \{a + (n-2)\, d\} + \cdots + a \qquad (8.6)$$

Adding together *8.5* and *8.6* term by term, in order gives

$$2S_n = [a + \{a + (n-1)\, d\}] + [(a+d) + \{a + (n-2)\, d\}]$$
$$+ [(a+2d) + \{a + (n-3)\, d\}] + \cdots + [\{a + (n-1)\, d\}] + a \quad (8.7)$$

i.e. $$2S_n = \{2a + (n-1)\, d\} + \{2a + (n-1)\, d\} + \cdots + \{2a + (n-1)\, d\}$$

that is, there are n terms of the form $2a + (n-1)\, d$.

Thus $$2S_n = n\{2a + (n-1)\, d\}$$

or $$S_n = \frac{n}{2} \{2a + (n-1)\, d\} \qquad (8.8)$$

Example 8.4
  In the series $5, 2, -1, \ldots$ what is
  (*a*) the 15th term
  (*b*) the sum of the first 15 terms, S_{15}?

(*a*) The first term, a, is 5; and the common difference, d, is -3. In the formula for the general term, $i = 15$. Hence the 15th term is $5 + (14)\,(-3) = -37$.
(*b*) Substituting in *8.8*, with $n = 15$, we have

$$S_{15} = \frac{15}{2} \{(2)\,(5) + (14)\,(-3)\}$$

$$= \frac{(15)\,(-32)}{2} = -\frac{480}{2} = -240$$

Example 8.5
Find the number of terms in the arithmetic series
(a) 8, 6, 4, ... and whose sum is 20;
(b) 5, 8, 11, ... and whose sum is 55.

(a) In this series, $a = 8$ and $d = -2$. From *8.8*, we have

$$20 = \frac{n}{2}\{16 + (n-1)(-2)\}$$

which reduces to the quadratic equation

$$n^2 - 9n + 20 = 0$$

i.e. $(n-4)(n-5) = 0$

Therefore $n = 4$ or $n = 5$

It can be seen that either answer is correct since the first five terms are 8, 6, 4, 2, 0; and so either the first four or five terms sum to 20.
(b) Here, $a = 5$ and $d = 3$. From *8.8*, we have

$$55 = \frac{n}{2}\{10 + (n-1)(3)\}$$

which reduces to $3n^2 + 7n - 110 = 0$

So $n = \dfrac{-7 \pm \sqrt{(49 + 1320)}}{6} \quad \dfrac{-7 \pm \sqrt{1369}}{6}$

$$= \frac{-7 \pm 37}{6} = \frac{30}{6} \quad \text{or} \quad -\frac{44}{6}$$

Now n, the number of terms in a series, must be a positive integer, and so the answer of $-44/6$ is inadmissible. Hence, the number of terms in the series is 5.

Σ- and Π-notations

Σ-notation

The sum of the first 4 terms of an arithmetic series whose first term is a, and whose common difference is d, is given by

$$S_4 = a + (a + d) + (a + 2d) + (a + 3d) \tag{8.9}$$

The individual terms on the right-hand side take the form of the general term $a + (i - 1) d$ with i changing from 1 to 4. This applies even to the first two terms, which could be represented as $a + 0d$ and $a + 1d$. The right-hand side of 8.9 can be written much more neatly, using the expression for the general term of the series, as

$$S_4 = \sum_{i=1}^{4} \{a + (i - 1) d\} \qquad (8.10)$$

The sign Σ means 'the sum of', and it implies the sum of terms of the type shown to the right of the sign, in the braces { } in this case. The actual terms to be summed are created by substituting integer values of i between the two limits shown below and above the Σ sign. If this is done in the *expression 8.10*, we obtain the summation on the right-hand side of *8.9*. Compare the Σ-notation with that of the definite integral, and read pages 121 to 122 of Chapter 7.

Example 8.6
(a) Express the following in Σ-notation:

$$x_1 + x_2 + x_3 + \cdots + x_n$$

(b) Expand $\Sigma_{i=3}^{7} ix^{(i-1)}$

(a) This is simply $\Sigma_{i=1}^{n} x_i$

(b) $\Sigma_{i=3}^{7} ix^{(i-1)} = 3x^{(3-1)} + 4x^{(4-1)} + 5x^{(5-1)} + 6x^{(6-1)} + 7x^{(7-1)}$
$\qquad = 3x^2 + 4x^3 + 5x^4 + 6x^5 + 7x^6$

Π-notation

Just as a summation can be concisely represented by means of a Σ-sign, so also can a product be similarly represented. Consider 4!; it is

$$4! = 1 \times 2 \times 3 \times 4$$

which can be written as $\qquad 4! = \prod_{i=1}^{4} i$

where the symbol Π means 'product of' the terms of type shown immediately to the right of the sign. The product notation is, however, very much less common than that for summation.

The geometric series

A *geometric series* is a series of terms that increases by a constant ratio, which may be positive or negative. The following are examples of geometric series:

$$2, 4, 8, 16, \ldots \qquad \text{common ratio} = +2$$
$$\tfrac{1}{2}, \tfrac{1}{4}, \tfrac{1}{8}, \tfrac{1}{16}, \ldots \qquad \text{common ratio} = +\tfrac{1}{2}$$
$$-\tfrac{1}{3}, 1, -3, 9, \ldots \qquad \text{common ratio} = -3$$
$$-2, \tfrac{1}{2}, -\tfrac{1}{8}, \tfrac{1}{32}, \ldots \qquad \text{common ratio} = -\tfrac{1}{4}$$

In a geometric series whose first term is a, and whose common ratio is r, the 2nd term is ar, the 3rd term is $(ar)\,r = ar^2$, etc.; hence the ith term is $ar^{(i-1)}$, and the last term is $ar^{(n-1)}$.

Sum of n terms

Consider a geometric series whose first term is a, and whose common ratio is r. In this series

$$S_n = a + ar + ar^2 + \cdots + ar^{(n-2)} + ar^{(n-1)} \qquad (8.11)$$

Multiply both sides by r:

$$rS_n = ar + ar^2 + \cdots + ar^{(n-2)} + ar^{(n-1)} + ar_n \qquad (8.12)$$

8.11 minus *8.12* gives

$$S_n - rS_n = a - ar^n$$

i.e. $\qquad S_n(1 - r) = a(1 - r^n)$

Hence $\qquad S_n = \dfrac{a(1 - r^n)}{1 - r} \qquad (8.13)$

Equally well, *8.11* could have been subtracted from *8.12* above, then the final result would have been

$$S_n = \frac{a(r^n - 1)}{r - 1} \qquad (8.14)$$

Expression *8.13* is used if the modulus of r (that is, the value of r regardless of sign) is less than 1, and *8.14* is used if the modulus of r is greater than 1. These results can be expressed concisely in the form of a theorem, as was done for the sum of an arithmetic series.

Theorem 8.2   The sum, S_n, of n terms of a geometric series, whose first term is a and whose common ratio is r, is given by

$$S_n = \frac{a(r^n - 1)}{r - 1} \quad |r| > 1$$

or by

$$S_n = \frac{a(1 - r^n)}{1 - r} \quad |r| < 1$$

Example 8.7
   In the series $3, -\frac{3}{2}, \frac{3}{4}, \ldots$; what is
   (a) the 13th term
   (b) the sum of the first 13 terms, S_{13}?

(a) The first term, a, is 3; and the common ratio, r, is $-\frac{1}{2}$. In the formula for the general term, $i = 13$. Hence the 13th term is

$$(3)\left(-\tfrac{1}{2}\right)^{12} = \tfrac{3}{4096}$$

(b) The modulus of r, $|r|$, is less than 1, so to sum the first 13 terms of the seriés, we employ *equation 8.13*:

$$S_{13} = \frac{3\{1 - (-\tfrac{1}{2})^{13}\}}{1 - (-\tfrac{1}{2})} = \frac{3\{1 + \tfrac{1}{8192}\}}{1 + \tfrac{1}{2}}$$

$$= \frac{(2)(8193)}{8192} .= 2.0002 \text{ (correct to 4 decimal places)}$$

Sum to infinity: divergence and convergence

The sum to infinity of a geometric series means the sum of an infinite number of terms of the series. Now at first sight this may seem a strange, even impossible, thing to find; and even if it were possible, the sum would apparently be infinitely large, and so useless for any practical purpose. The last two statements of the previous sentence are true enough for the arithmetic series, but they are not necessarily true for the geometric series. In fact, the concept of the sum to infinity of a series is very important in mathematics, and we shall use the geometric series as an illustration.

   First, consider the geometric series: 2, 4, 8, 16, ..., in which the common ratio is 2. The sum of the first two terms is 6, of the first three 14, of the first four 30, and of the first five terms 62. Clearly, not only is S_n increasing as n

increases, but the increase of s_n with each successive increment in n becomes steadily greater. Evidently, if an infinite number of terms were added together the sum would become infinitely large. Thus for the geometric series $2, 4, 8, 16, \ldots$ we may write

$$S_n \to \infty, \quad \text{as} \quad n \to \infty$$

Such a series is said to be **divergent**.

Now consider the series $1, \frac{1}{2}, \frac{1}{4}, \frac{1}{8}, \ldots$ The sum of the first two terms is $1\frac{1}{2}$, of the first three $1\frac{3}{4}$, of the first four $1\frac{7}{8}$, and of the first five terms $1\frac{15}{16}$. In this case, although S_n is increasing with each term added, it seems to be doing so more and more slowly. We are thus led to ask for this series, 'is there a limit above which S_n cannot rise, no matter how large n is?' This question may be answered by the use of the formula for the sum of a geometric series; *expression 8.13*, since $r = \frac{1}{2}$ for this series, and $a = 1$. Substituting in *8.13*, we have for the sum of n terms

$$S_n = \frac{1\{1 - (\frac{1}{2})^n\}}{1 - \frac{1}{2}} = 2\{1 - (\frac{1}{2})^n\} \tag{8.15}$$

Consider now what happens when n gets very large, in other words $\frac{1}{2}$ is raised to a large power. Recall that $(\frac{1}{2})^2 = \frac{1}{4}$, $(\frac{1}{2})^3 = \frac{1}{8}$, $(\frac{1}{2})^4 = \frac{1}{16}$, and so on. The larger n becomes, the smaller is $(\frac{1}{2})^n$; so that as

$$n \to \infty, \quad (\frac{1}{2})^n \to 0$$

Hence, from *8.15*, we can write that as

$$n \to \infty, \quad S_n \to 2$$

Thus the sum to infinity of the series $1, \frac{1}{2}, \frac{1}{4}, \frac{1}{8}, \ldots$ is 2, a perfectly ordinary and usable number. The series is said to be **convergent** because, as more and more terms are added in, the value of the sum of the terms converges to a definite limit.

For a geometric series, the criterion for convergence or divergence is the value of the common ratio, r. If the modulus of the common ratio, $|r|$, is greater than unity, then the series is divergent, and if $|r|$ is less than unity, the series is convergent.

Theorem 8.3 The sum to infinity of a congruent geometric series is given by

$$S_\infty = \frac{a}{1 - r} \tag{8.16}$$

Since this series is stated to be convergent, which implies $|r| < 1$, S_∞ can be obtained from *formula 8.13*. It is immediately apparent that with $|r| < 1$ as $n \to \infty$, $r^n \to 0$. Hence, the result *8.16* is proved.

Example 8.8
Sum to infinity the series 3, $-\frac{3}{4}$, $\frac{3}{16}$, ... In this series, $a = 3$ and $r = -\frac{1}{4}$. So, substituting into *formula 8.16*

$$S_\infty = \frac{3}{1 - (-\frac{1}{4})} = \frac{12}{5} = 2.4$$

Example 8.9
Convert the recurring decimal $15.2\dot{6}\dot{4}$ into a (non-decimal) fraction (vulgar fraction).

$$15.2\dot{6}\dot{4} = 15 + \frac{2}{10} + \frac{64}{10^3} + \frac{64}{10^5} + \frac{64}{10^7} + \cdots$$

$$= 15 + \frac{2}{10} + \frac{64}{10^3} \left\{ 1 + \frac{1}{10^2} + \frac{1}{10^4} + \cdots \right\}$$

$$= 15 + \frac{2}{10} + \frac{64}{10^3} \left\{ \frac{1}{1 - (\frac{1}{10})^2} \right\}$$

(summing the series in the braces, above, to infinity)

$$= 15 + \tfrac{2}{10} + \tfrac{64}{990} = 15\tfrac{262}{990}$$

At first sight, series may appear to be of intrinsic mathematical interest only, but *example 8.9* demonstrates a very practical application of the geometric series. Most series have practical applications in one way or another, which make them important to the user of mathematical methods.

The binomial series

By direct multiplication, we can show that

$$(a + x)^2 = a^2 + 2ax + x^2$$

$$(a + x)^3 = a^3 + 3a^2 x + 3ax^2 + x^3$$

$$(a + x)^4 = a^4 + 4a^3 x + 6a^2 x^2 + 4ax^3 + x^4$$

and so on. These results can be generalized, and also extended to cover situations where the exponent, n, in the expression $(a + x)^n$ is any real number, by a result known as the **binomial theorem**. This theorem gives the following result:

$$(a + x)^n = a^n + na^{(n-1)} x + \frac{n(n-1)}{2!} a^{(n-2)} x^2 + \frac{n(n-1)(n-2)}{3!} a^{(n-3)} x^3$$

$$+ \cdots + \frac{n(n-1)(n-2)\ldots(n-i+2)}{(i-1)!} a^{(n-i+1)} x^{(i-1)} + \cdots$$

$$(8.17)$$

This relationship can be written more succinctly by a fuller use of the factorial notation:

$$(a + x)^n = a^n + na^{(n-1)} x + \frac{n!}{2!(n-2)!} a^{(n-2)} x^2 + \frac{n!}{3!(n-3)!} a^{(n-3)} x^3$$

$$+ \cdots + \frac{n!}{(i-1)!(n-i+1)!} a^{(n-i+1)} x^{(i-1)} + \cdots \qquad (8.18)$$

The last term written out in *8.17* and *8.18* is the ith term of the series, and the sum of the series on the right-hand side of these two expressions is, of course, the left-hand side, $(a + x)^n$. This is the reverse situation to that considered hitherto. When discussing the arithmetic and geometric series, we already had the series of terms and were interested in finding a concise formula for the sum of the terms. Now the question is posed in reverse: can we express the concise expression $(a + x)^n$ as a series of terms? *Relationships 8.17* and *8.18* answer this question in the affirmative, but the general proof of the binomial theorem is outside the scope of this book.

Example 8.10
Expand

(a) $(a + x)^5$ (b) $(a - x)^4$ (c) $(2 - 3y)^4$

(a) This involves the direct use of *8.17*:

$$(a + x)^5 = a^5 + 5a^4 x + \frac{(5)(4)}{(1)(2)} a^3 x^2 + \frac{(5)(4)(3)}{(1)(2)(3)} a^2 x^3$$

$$+ \frac{(5)(4)(3)(2)}{(1)(2)(3)(4)} ax^4 + \frac{(5)(4)(3)(2)(1)}{(1)(2)(3)(4)(5)} x^5$$

$$(a + x)^5 = a^5 + 5a^4 x + 10a^3 x^2 + 10a^2 x^3 + 5ax^4 + x^5$$

(b) Put x in 8.17 equal to $-x$. Then 8.17 becomes, with $n = 4$,

$$\{a + (-x)\}^4 = a^4 + 4a^3(-x) + \frac{(4)(3)}{(1)(2)} a^2(-x)^2 + \frac{(4)(3)(2)}{(1)(2)(3)} a(-x)^3$$

$$+ \frac{(4)(3)(2)(1)}{(1)(2)(3)(4)} (-x)^4$$

i.e. $\quad (a - x)^4 = a^4 - 4a^3 x + 6a^2 x^2 - 4ax^3 + x^4$

(c) In 8.17, put x equal to $-3y$, $a = 2$, and $n = 4$,

then $\quad \{2 + (-3y)\}^4 = (2)^4 + (4)(2)^3(-3y) + \frac{(4)(3)}{(1)(2)} (2)^2 (-3y)^2$

$$+ \frac{(4)(3)(2)}{(1)(2)(3)} (2)(-3y)^3 + \frac{(4)(3)(2)(1)}{(1)(2)(3)(4)} (-3y)^4$$

$$= 16 + 32(-3y) + 24(-3y)^2 + 8(-3y)^3 + (-3y)^4$$

i.e. $\quad (2 - 3y)^4 = 16 - 96y + 216y^2 - 216y^3 + 81y^4$

In the above examples where the power, n, is a positive integer, it should be noted that there are $n + 1$ terms in the expansion. In the expansion of $(a + x)^n$, the first term is a^n (i.e. $a^n x^0$), and the last term is x^n (i.e. $a^0 x^n$); also, in each term the sum of the powers of a and x is always equal to n.

From a biological viewpoint, the chief application of the binomial series occurs in a statistical distribution, known as the binomial distribution, which arises particularly in genetics and animal behaviour.

EXERCISES

1. Solve (general solution) the following differential equations, in which the variables are separable.
(a) $dy/dx = 1 - y/a$ (b) $x(x + 1) dy/dx = y - 1$
(c) $2x \, dy/dx = 1 - y^2$

2. Find the arithmetic series whose third term is -11, and whose seventh term is 5. What is the twentieth term of the series, and what is the sum of the first twenty terms?

3. Expand

(a) $\sum\limits_{i=1}^{5} x_i^2$ (b) $a \sum\limits_{i=0}^{3} x^i$ (c) $\sum\limits_{i=1}^{4} \frac{\sqrt{x_i}}{a + x_i}$ (d) $\prod\limits_{i=1}^{4} x_i$

4. Express the following in Σ- or Π-notation, as appropriate:
(a) $(x_1 + x_2 + x_3)/3$ (b) $2x^5 + x^4 - x^2 - 2x$ (c) $(x_1 x_2 x_3)/3$
(d) $\sqrt[3]{(x_1 x_2 x_3)}$

5. The sum to infinity of a geometric series is 36, and the second term of the series is 8. What is the series?

6. Convert to vulgar fractions:
(a) $0.\dot{1}$ (b) $0.\dot{2}\dot{7}$ (c) $0.1\dot{2}\dot{3}$

7. Expand
(a) $(x + 1)^5$ (b) $(x - 1)^5$ (c) $(3x - 2y)^4$ (d) $(x^2 + 3y^2)^4$

9

The exponential and related functions of biological importance

The exponential function

In the discussion of the growth of a microbial culture (Chapter 6), a primary concept was that of the rate of increase of the number of organisms in an unlimited environment. Since under these conditions it is considered that each cell, on average, divides at a constant rate, then the rate of increase of the entire culture at any instant of time, t, will be proportional to the actual number of cells, n, present at that instant;

$$\text{i.e.} \quad \frac{dn}{dt} = kn \qquad (9.1)$$

where k is the constant of proportionality. Because 9.1 contains the term dn/dt, the expression is a differential equation, and we have already shown (page 135) that a differential equation of the form of 9.1 can be 'solved' to obtain a function of the form

$$n = f(t) \qquad (9.2)$$

which gives the number of organisms present at time t.

If *equations 9.1* and *9.2* are cast into more general x–y terms, we have

$$\frac{dy}{dx} = ky \qquad (9.3)$$

$$\text{and} \quad y = f(x) \qquad (9.4)$$

Now 9.4 is a generalized way of writing down functions, and includes those that we have already studied, such as polynomials; evidently the form of equation in 9.3, which we have hitherto met only in connexion with our culture growth problems, can give rise to, or can be obtained from, *equation 9.4*.

Now put $k = 1$ in 9.3; then we have

$$\frac{dy}{dx} = y \qquad (9.5)$$

In words, 'the rate of change of y with respect to x numerically equals y'. If we substitute for y in 9.5 using 9.4, we have

$$\frac{dy}{dx} = f(x) \qquad (9.6)$$

So we are implying that there is a function of the form $y = f(x)$ whose derivative is the same as the function itself, i.e. that $dy/dx = f(x)$. Put in another way, we are saying that for this particular function of x, the rate of change of $f(x)$ with respect to x is always numerically equal to $f(x)$ for any specified value of x. Precisely what form does $f(x)$ take for it to have this property?

Consider the third degree polynomial function $y = x^3$, where $dy/dx = 3x^2$. Draw-up a small table of values of y and of dy/dx for a few values of x:

x	y	$\dfrac{dy}{dx}$
0	0	0
1	1	3
2	8	12
3	27	27
4	64	48

We see that for this function in the range of x selected, y and dy/dx are equal at two values of x, namely $x = 0$ and $x = 3$. Between these two values of x the gradient of the curve $y = x^3$ exceeds the value of y itself, i.e. $dy/dx > y$, and when $x > 3$ the gradient is less than y, i.e. $dy/dx < y$. It can be stated now that no polynomial function can fulfil our present requirements that $dy/dx = y$ *always*.

The number e and the exponential series

Let us turn our attention to another function, namely $y = ac^x$, where a and c are defined to be *positive* constants. It can be shown that $dy/dx = ac^x \log_e c$. Clearly, the derivative is rather similar to the original, differing only by the inclusion of a constant multiplier, $\log_e c$. Now the logarithm of a number to its own base always equals 1, e.g. $\log_{10} 10 = 1$ (see Exercise 5(a) at the end of Chapter 2); so that if $c = e$, we have $y = a\,e^x$, and $dy/dx = a\,e^x \log_e e = a\,e^x$. So, what is the value of e?

As already remarked, the fundamental property of e is that if we consider the function $y = e^x$, then dy/dx also equals e^x. Dispensing with the variable y, we can write

$$\frac{d(e^x)}{dx} = e^x \qquad (9.7)$$

Relationship 9.7 can be said to define e^x, and hence e by putting $x = 1$, but in itself does not give us the numerical value of e. What we now ask is, can we find some more elementary function of x which can be differentiated from first principles, and when so differentiated it remains unchanged? If we can find such a function of x, then it must equal e^x.

The exponential series

Consider the series of terms

$$1 + x + \frac{x^2}{2!} + \frac{x^3}{3!} + \frac{x^4}{4!} + \cdots + \frac{x^i}{i!} + \cdots \tag{9.8}$$

This series is convergent (page 143) because in successive terms $i!$ increases faster than x^i. If x is small convergence is rapid, but if x is large the series converges slowly; but the main point is that the further to the right one goes in *9.8*, the smaller the terms become. Although the first two terms of the series appear to be different from the remainder, we can bring them into line by writing the second term, x, as $x^1/1!$ and the first term, 1, as $x^0/0!$ (remembering that any number raised to the power zero is 1, and that $0! = 1$; page 9). So the series may be written succinctly as

$$\sum_{i=0}^{\infty} \frac{x^i}{i!}$$

implying the sum to infinity of the series.

Now consider the differentiation of *9.8*. Close inspection shows the function to be a polynomial (of infinite degree), so we can differentiate it term by term. Let us put the whole series equal to $f(x)$, as clearly *9.8* is a function of x, despite being written as a series. Writing out the first few terms, we have

$$f(x) = 1 + x + \frac{x^2}{2!} + \frac{x^3}{3!} + \frac{x^4}{4!} + \frac{x^5}{5!} + \cdots \tag{9.9}$$

then $$f'(x) = 1 + \frac{2x}{2!} + \frac{3x^2}{3!} + \frac{4x^3}{4!} + \frac{5x^4}{5!} + \cdots$$

and on cancelling in the numerator and denominator of each term, we find

$$f'(x) = 1 + x + \frac{x^2}{2!} + \frac{x^3}{3!} + \frac{x^4}{4!} + \cdots \tag{9.10}$$

You will notice that the *series 9.9* and *9.10* are exactly the same. True, all the terms have been shifted one place to the right, but because the series is infinite, the final result is unchanged. Hence $f(x) = f'(x)$, and so the series must be equal to e^x. Therefore

$$e^x = 1 + x + \frac{x^2}{2!} + \frac{x^3}{3!} + \cdots = \sum_{i=0}^{\infty} \frac{x^i}{i!} \qquad (9.11)$$

and so both *relationships 9.7* and *9.11* define e^x. An alternative way of writing e^x is exp (x).

The value of e

Now it is quite simple to evaluate e. All we need to do is to put $x = 1$ in *9.11*:

$$e = 1 + 1 + \frac{1}{2!} + \frac{1}{3!} + \cdots = \sum_{i=1}^{\infty} \frac{1}{(i-1)!} \qquad (9.12)$$

Consider the first 10 terms, in the following table:

i	$i-1$	$(i-1)!$	$1/(i-1)!$
1	0	1	1.000 000
2	1	1	1.000 000
3	2	2	0.500 000
4	3	6	0.166 667
5	4	24	0.041 667
6	5	120	0.008 333
7	6	720	0.001 389
8	7	5 040	0.000 189
9	8	40 320	0.000 025
10	9	362 880	0.000 003

$$\sum_{i=1}^{10} \frac{1}{(i-1)!} = 2.718\ 282$$

Hence, by considering the first 10 terms in the exponential series, we find that $e = 2.71828$, correct to 5 decimal places. If a more accurate evaluation of e were required, that is, correct to more decimal places, all that need be done is to add in more terms to the series. By now it should be apparent why e is irrational, since it is equal to the sum to infinity of a series of terms.

The exponential function

If
$$y = e^x \tag{9.13}$$

the curve of this function will have an intercept on the y-axis of 1. More generally, if $y = ae^x$ the intercept is a. We also know that

$$\frac{dy}{dx} = ae^x \quad \text{or} \quad \frac{dy}{dx} = y \tag{9.14}$$

The second equation in *9.14* is the premise with which we started, namely, 'that the rate of change of y with respect to x is equal to y itself'. Now it would be very rare to find that the rate of change would actually *equal* y, but instead it would be proportional

$$\text{i.e.} \quad \frac{dy}{dx} = by$$

where b is the constant of proportionality. If this is the case the first equation of *9.14* becomes

$$y = ae^{bx} \tag{9.15}$$

and *9.15* is called the **exponential function**. Strictly, $y = e^x$ is *the* exponential function, but *9.15* is a more general, and useful form.

If we take natural logarithms of both sides of *9.15*, we have

$$\log_e y = \log_e a + bx \log_e e$$

and since $\log_e e = 1$, the above becomes

$$\log_e y = \log_e a + bx \tag{9.16}$$

Equation 9.16 represents a linear relationship between $\log_e y$ and x with a gradient of b and an intercept of $\log_e a$. Graphs are shown in Fig. 9.1 of *equation 9.15*, (a) and (b), and of *equation 9.16*, (c) and (d). In (a) and (c) are shown the curve and line when b takes positive values, and in (b) and (d) those when b is negative.

The exponential function is widely used in biology. With a positive value of b, the use of the function is confined to describing a particular form of growth (page 160), but the function in which b is negative has wider application. The decay of a radioactive element with time (of outstanding importance to the biologist interested in physiology and metabolism) proceeds according to the exponential function, and this function also describes the way that the intensity of a beam of light is attenuated as it passes through a liquid (Beer's Law). Empirically, the exponential function may also describe the attenuation of light

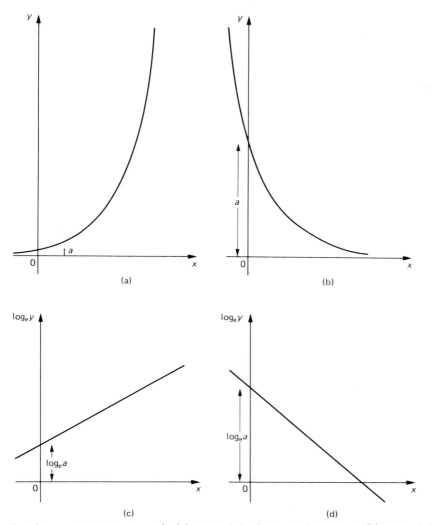

Fig. 9.1 The function $y = ae^{bx}$: (**a**) curve of the function when $b > 1$, (**b**) curve of the function when $b < 1$, (**c**) the straight line $\log_e y = \log_e a + bx$ corresponding to (a), (**d**) the straight line corresponding to (b).

through a foliage canopy and also the natural decline in numbers of individual plants in a cohort from seedling emergence to maturity. It is a matter of common observation that if a large number of seedlings of a particular species are found on a piece of bare ground, only a small proportion of these will reach maturity, and the decline in number of individuals with time may often be described by the exponential function. The example below illustrates this idea, as well as showing how the function is manipulated in calculations.

Example 9.1

In a cohort of foxglove (*Digitalis purpurea*) plants, the number, y, surviving at time t (measured in months from the emergence of the seedlings) was found to conform to the equation

$$y = 100\,e^{-0.2310t}$$

(*a*) What was the original number of seedlings in the cohort?

(*b*) What was the half-life of the cohort, i.e. the time when half the original number of individuals had died?

(*c*) The foxglove is a biennial, germinating in spring and flowering in the summer of the following year. Assuming that 15 months are required for vegetative growth, how many individuals of the cohort would be likely to survive to flower?

(*a*) Since time is measured from seedling emergence, we put $t = 0$; then, from the equation, $y = 100$. So the original number of seedlings was 100.

(*b*) To calculate the time at which half the number of seedlings have died, we substitute the number surviving into the equation for y and solve for t. In this case, half the original number surviving gives $y = 50$, and so

$$50 = 100\,e^{-0.2130t}$$

or $$0.5 = e^{-0.2310t}$$

Taking the reciprocal of both sides gives $e^{0.2310t} = 2$

Take natural logarithms of both sides $0.2310t = \log_e 2$

Now $\log_e 2 = 0.6931$

and so $t = \dfrac{0.6931}{0.2310} \simeq 3$

Therefore 50% of the plants die during the first three months from seedling emergence.

(*c*) As we are told that 15 months are required for vegetative growth, we merely substitute $t = 15$ into the equation and solve for y.

$$y = 100\,e^{-0.2310 \times 15}$$

i.e. $$y = 100\,e^{-3.465}$$

or $$y = 100/e^{3.465}$$

Now $e^{3.465} = 31.98$, and so $y \simeq 3.127$

Hence, three plants are likely to survive to flower.

It should be noted that an exponential function, such as the one shown in the above example, would be fitted to data in its linear form, *equation 9.16*. In the foxglove cohort, observations on the number of individuals surviving would be made from time to time, and then a graph would be plotted of $\log_e$ (number of individuals) against time. The points should lie roughly on a straight line, and such a line may then be fitted in the way mentioned in Chapter 3 (page 40) and values of the constants $\log_e a$ (and hence a) and b obtained.

The logarithmic function

In the simplified form of the exponential function, where both the constants a and b are unity, we have $y = e^x$. Referring to *equations 2.12* and *2.13*, and the definition of a logarithm (page 12), we see that this implies

$$x = \log_e y \qquad (9.17)$$

If we now reverse the roles of x and y, *9.17* becomes

$$y = \log_e x$$

A more generalized form of this is

$$y = \log_e (ax^b) \qquad (9.18)$$

Expanding the right-hand side gives

$$y = \log_e a + \log_e x^b$$

$$\text{or} \qquad y = \log_e a + b \log_e x \qquad (9.19)$$

Relationship 9.19 shows y to be a linear function of $\log_e x$, the reverse situation to the exponential function in which $\log_e y$ is a linear function of x.

Graphs of the **logarithmic function** are shown in Fig. 9.2(a) and (b), illustrating the ordinary form (*equation 9.18*), and (c) and (d), illustrating the linear form (*equation 9.19*). In (a) and (c) of Fig. 9.2, b is positive, while in (b) and (d) this constant is negative.

The logarithmic function may describe empirically the relationship between the concentration, x, of a substance having a specific physiological action on an organism (e.g. a plant growth hormone) and a measurement, y, of the response of the organism (e.g. growth made) to the substance. If the logarithm of the concentration of the applied substance is plotted against the measurement of the response of the organism, and an approximately linear distribution of points is obtained, then the logarithmic function adequately describes the data, and the values of the constants a and b may be estimated. It is then possible to assess the concentration at which no response occurs, by observing the intercept of the curve *on the x-axis*, which is given by

$$x_0 = e^{-(\log_e a)/b} \qquad \text{(see Fig. 9.2a).}$$

The calculus of exponential and logarithmic functions

Basic results

Since $d(e^x)/dx = e^x$, then $\int e^x\,dx = e^x + c$. For the simple logarithmic function $y = \log_e x$, using the definition of a logarithm,

$$x = e^y \tag{9.20}$$

$$\text{Now} \qquad \frac{dx}{dy} = e^y$$

$$\text{and so} \qquad \frac{dy}{dx} = \frac{1}{e^y}$$

But $e^y = x$, *9.20*, so

$$\frac{dy}{dx} = \frac{1}{x}$$

Hence $d(\log_e x)/dx = 1/x$. This means that $\int dx/x = \log_e x + c$. These are very important results, and are summarized below:

$$\frac{d(e^x)}{dx} = e^x \tag{9.21}$$

$$\frac{d(\log_e x)}{dx} = \frac{1}{x} \tag{9.22}$$

$$\int e^x\,dx = e^x + c \tag{9.23}$$

$$\int \frac{dx}{x} = \log_e x + c \tag{9.24}$$

Equations 9.21 and *9.22* are additional basic rules for differentiation, while *9.23* and *9.24* are standard integrals.

It will be recalled that the standard integral first introduced, equation *7.6* on page 105, was not valid for $n = -1$, but was valid for any other value of n. The standard integral *9.24* covers the case where $n = -1$;

$$\text{i.e.} \qquad \int x^{-1}\,dx = \log_e x + c$$

Some more complicated functions

Consider the exponential function, $y = e^{bx}$, i.e. where the coefficient a (*equation 9.15*) is 1. This can be differentiated by the function of a function rule. Put

$$u = bx \qquad \text{then} \qquad y = e^u$$

$$\frac{du}{dx} = b \qquad \text{and} \qquad \frac{dy}{du} = e^u$$

So
$$\frac{dy}{dx} = \frac{dy}{du} \cdot \frac{du}{dx} = b\, e^u = b\, e^{bx}$$

Hence
$$\frac{d(e^{bx})}{dx} = b\, e^{bx} \qquad\qquad (9.25)$$

From *9.25* we can say that

$$\int b\, e^{bx}\, dx = e^{bx} + c$$

i.e.
$$b \int e^{bx}\, dx = e^{bx} + c$$

Dividing both sides by b, we have

$$\int e^{bx}\, dx = \frac{1}{b} e^{bx} + k \qquad\qquad (9.26)$$

where $k = c/b$. Thus *expressions 9.25* and *9.26* give the derivative and integral, respectively, of the function e^{bx}.

Now consider a different form of the logarithmic function,

$$y = \log_e (a + bx)$$

Again, this may be differentiated by the function of a function rule.

Put $\qquad u = a + bx \qquad$ then $\qquad y = \log_e u$

$$\frac{du}{dx} = b \qquad \text{and} \qquad \frac{dy}{du} = \frac{1}{u}$$

So
$$\frac{dy}{dx} = \frac{dy}{du} \cdot \frac{du}{dx} = b\,\frac{1}{u} = \frac{b}{a + bx}$$

Hence
$$\frac{d\{\log_e (a + bx)\}}{dx} = \frac{b}{a + bx} \qquad\qquad (9.27)$$

From *9.27* we can say that

$$\int \frac{b \, dx}{a + bx} = \log_e (a + bx) + c$$

i.e. $\qquad b \int \frac{dx}{a + bx} = \log_e (a + bx) + c$

Dividing both sides by b, we have

$$\int \frac{dx}{a + bx} = \frac{1}{b} \log_e (a + bx) + k \qquad\qquad (9.28)$$

where $k = c/b$. Thus *expressions 9.27* and *9.28* give the derivative and integral of $\log_e (a + bx)$ and $1/(a + bx)$, respectively.

Example 9.2
   Find (a) $d\{e^{(2+3x)}\}/dx$ (b) $\int dx/(2 - 5x)$

(a) Use the function of a function rule.

$\qquad$ Put $\qquad y = e^{(2+3x)}$ $\qquad$ and $\qquad u = 2 + 3x,$ $\qquad$ then $\qquad y = e^u$

$$\frac{du}{dx} = 3 \qquad \text{and} \qquad \frac{dy}{du} = e^u$$

$\qquad$ So $\qquad \dfrac{dy}{dx} = \dfrac{dy}{du} \cdot \dfrac{du}{dx} = 3e^u = 3e^{(2+3x)}$

$\qquad\qquad$ Hence $\qquad d\{e^{(2+3x)}\}/dx = 3e^{(2+3x)}$

(b) This is a direct application of the standard integral, *9.28*, where $a = 2$ and $b = -5$. Hence

$$\int \frac{dx}{2 - 5x} = -\frac{1}{5} \log_e (2 - 5x) + c)$$

Growth functions

One of the areas of biology to which mathematical analysis has usefully been employed is in the analysis of growth of cells, whole organisms or populations. We have already used an example applying calculus to a biological problem, namely the growth of a microbial culture. In that example, we formulated, in mathematical terms, our initial knowledge of the way in which the rate of increase of cell number should change in relation to the number of cells actually present. We arrived at an equation summarizing this knowledge, *6.12*. Because this equation contains a derivative, it is called a differential equation, and it relates the rate of increase in cell number, *n*, to cell number present at time *t*:

$$\frac{dn}{dt} = f(n) \qquad (9.29)$$

We have also shown (Chapter 8) that from this differential equation we can obtain another equation giving a relationship between cell number and time,

$$\text{i.e.} \qquad n = F(t) \qquad (9.30)$$

If it is assumed that the microbial colony completely obeyed the assumptions we made, and therefore also the equations derived for it, the number of cells present at any selected time *t* is given by *equation 9.30*; and the curve of *9.30* would appear as in Fig. 6.5(b). An equation of the form of *9.30* is called a **growth function**, and its curve may be known as a **growth curve**; but the latter term is often applied to a plot of actual observations as shown in Fig. 6.5(a), especially if adjacent points are joined by straight lines.

Now although the differential *equation 6.12* is simple in form, that is *f(n)* in *9.29* is simple, the resulting *F(t)*, *9.30*, is not an equally elementary function (see page 132). So first we shall consider an even simpler differential equation giving rise to a growth function. Since quantities other than cell number may be used for describing growth, we shall use *y* instead of *n* in what follows.

The exponential function

We have remarked that, when the number of cells in a microbial culture is small relative to the quantity of medium in which the colony finds itself, the rate of growth at any instant of time *t* would be proportional to the size of the colony (number of organisms present), *y*, at that time (page 91). This could be put into mathematical notation as

$$\frac{dy}{dt} = by \qquad (9.31)$$

On page 152, it was indicated that when the differential *equation 9.31* is solved to give *y* as a function of time, the result is the exponential function

$$y = a\, e^{bt} \qquad (9.32)$$

where $a = e^c$, *c* being the constant of integration. Since *9.32* is an exponential function, any biological material, from a cell to a whole population of multicellular organisms, which is growing in such a manner that it can be described by *equation 9.32* is said to be **growing exponentially**.

For any particular set of growth data, how can we establish whether it conforms to the exponential function or not? To do this rigorously involves concepts in statistics, and is therefore outside the scope of this book; but one can get some idea from a graphical examination of the data. However, it would not be very illuminating to plot the growth measurements, *y*, obtained at a number of times, *t*, on a *t–y* graph and try to see whether, assuming there was no sample to sample variability (page 90), the data lay on an exponential curve. Even if there were no sampling variability, it would be impossible to assert that the data followed the curve of an exponential function rather than that of one of many other types of function. The only type of curve that it is possible to distinguish with any degree of certainty is a straight line. In the case of the exponential function, we know that the relationship between $\log_e y$ and *t* is linear, *9.16*; hence, if the logarithms of our growth measurements are plotted against time, as in Fig. 9.1(c), and if the resulting points look as though they would lie on a straight line if sampling variability were absent, then it can be inferred that exponential growth underlies the data produced.

Now although growth cannot be exponential over long periods, such a pattern is found in many biological situations over short periods of time. Thus growth is exponential, or approximately so, in the early stages of growth of a microbial culture. In higher plants, growth of the seedling for a while after it has become independent of seed reserves is often said to be exponential. As far as animals are concerned, it seems that exponential growth only occurs in the embryonic stage.

Absolute and relative growth rates

It may be asked just what form does a growth function and its curve take if it is not exponential? To answer this, we need to return to the growth rate form of the exponential function, *equation 9.31*. Now in biology, the growth rate as defined by dy/dt is often not a very useful concept, and another form of growth rate is usually more meaningful. In order to distinguish between the two types of growth rate, they are given different names; the growth rate shown in *equation 9.31* as dy/dt is called the **absolute growth rate**. So if growth is exponential, this means that at any instant of time the absolute growth rate is proportional to size already attained. The last sentence indicates why the absolute growth rate is not very helpful to the growth analyst: young sunflower plants have a higher absolute growth rate than plants of chickweed at the same stage of development, and most of the differences in absolute growth rate between the plants of these two species is due to the size difference, rather than to any difference in the constants of proportionality, *b*.

Now if both sides of *9.31* are divided by *y* (the size), we have

$$\frac{1}{y} \cdot \frac{dy}{dt} = b \qquad (9.33)$$

which provides a new definition of rate of growth. The quantity on the left-hand side of *9.33* is called the **relative growth rate**, or sometimes the **specific growth rate**; it is the absolute growth rate divided by size, or the absolute growth rate per unit size. Viewed in this way, it is evident that the *relative growth rate is a measure of the efficiency of the material in producing new material*.

When growth is exponential, we see from *9.33* that this efficiency is constant; in particular, it does not change with time. If relative growth rate is increasing with time, we could describe this as supra-exponential growth. This type of growth occurs only over very limited periods in well-defined circumstances, e.g. when a new bacterial culture is changing from its lag phase to its exponential phase, or in a very young seedling when photosynthesis is rapidly increasing and overcoming the loss in dry weight due to respiration. It is always important when considering growth to state what aspect of growth is being studied. In the above examples, it is cell number in the microbial culture, while growth in higher plants is normally measured in terms of dry weight.

After possibly going through a period of exponential growth, an organism continues growing but with a *declining relative growth rate*. For many, or perhaps most, organisms all their growth is made in this way, and so to describe such growth mathematically we need to look for a function of the form $y = f(t)$ in which $(1/y)\,(dy/dt)$ is continuously decreasing.

Before leaving this section, and examining two functions with the above property, it should be noted that relative growth rate is often given its own symbol, R, and so

$$R = \frac{1}{y} \cdot \frac{dy}{dt} \qquad (9.34)$$

When the exponential function is used to describe growth, i.e. *equation 9.32*, we see from *equations 9.33* and *9.34* that $R = b$; so the exponential growth function may be written as

$$y = a\,e^{Rt} \qquad (9.35)$$

which emphasizes the fact that when growth is exponential, relative growth rate is constant.

The logistic function

Reverting yet again to our old friend the microbial culture, we saw that the differential *equation 6.12* described the absolute growth rate of the culture from (in theory) no individuals to (again in theory) the maximum number of individual cells the culture could contain. Thus the equation is an overall description of events throughout the growing life of the colony. If we divide both sides of *6.12* by y (using the more general symbol y instead of n), we have

$$\frac{1}{y} \cdot \frac{dy}{dt} = k \left(1 - \frac{y}{a}\right)$$

i.e. $\qquad R = k - \frac{k}{a} y \qquad (9.36)$

Relationship 9.36 shows the relative growth rate to be a linear function of size, with gradient $-k/a$ and intercept k (Fig. 9.3). Evidently, relative growth rate is nowhere constant, but declines continuously (in fact, linearly) with increasing size. The growth function obtained by solving the differential equation is

$$y = \frac{a}{1 + b\,e^{-kt}} \qquad (9.37)$$

where $b = e^c$, c being the constant of integration (page 132).

We already know that the maximum size is a, but this fact may also be deduced directly from *equation 9.37*. As $t \to \infty$, $e^{kt} \to \infty$; hence $e^{-kt} \to 0$. Thus $b\,e^{-kt} \to 0$, and so $(1 + b\,e^{-kt}) \to 1$. So we see that as $t \to \infty$, $y \to a$; the curve is asymptotic to the maximum size a. The intercept on the y-axis is found by putting $t = 0$ in *9.37*, we can call it y_0. When $t = 0$, $e^{-kt} = 1$, and so $b\,e^{-kt} = b$. Thus

$$y_0 = \frac{a}{1 + b} \qquad (9.38)$$

The function *9.37* has various names, the two commonest being the **logistic** and the **autocatalytic**. The curve of the function is shown in Fig. 9.4(a), and it can be seen that the gradient (the absolute growth rate) is positive throughout.

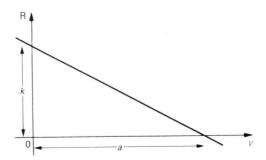

Fig. 9.3 The straight line $R = k - ky/a$.

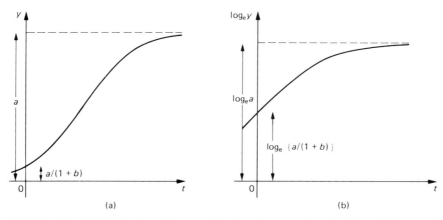

(a)   (b)

Fig. 9.4   The logistic function: (**a**) curve of the function $y = a/(1 + b\,e^{-kt})$, and (**b**) curve of the function $\log_e y = \log_e a - \log_e(1 + b\,e^{-kt})$.

The gradient increases in value near the beginning and decreases towards the end of the growth period so there is a point of inflexion: this occurs at $y = a/2$, that is, when half the total growth has been made.

Logarithmic form

For a number of reasons, it is most convenient to deal with growth data in logarithmic form. One of these reasons is that a test for exponential growth can be made (page 161) when the data are in this form; another reason is that growth functions have to be fitted to data in logarithmic form. The equation of the logarithmic form of the logistic function can be obtained by taking natural logarithms of both sides of 9.37 to give:

$$\log_e y = \log_e a - \log_e (1 + b\,e^{-kt}) \qquad (9.39)$$

The curve for this is shown in Fig. 9.4(b). There is no point of inflexion, and the gradient, though always positive, declines throughout. The intercept is now $\log_e\{a/(1 + b)\}$, which could be negative, while the asymptote is, of course, $\log_e a$.

Linear form

A linear form of the logistic function can be obtained by first rearranging 9.37 to give

$$\frac{a}{y} = 1 + b\,e^{-kt}$$

and so

$$\frac{a}{y} - 1 = b\,e^{-kt}$$

Taking natural logarithms of both sides gives

$$\log_e \left(\frac{a}{y} - 1 \right) = \log_e b - kt \qquad (9.40)$$

Thus $\log_e (a/y - 1)$ is a linear function of t, with a gradient of $-k$ and an intercept of $\log_e b$.

Relationship 9.40 could be used to test whether a set of data reasonably conforms to a logistic function, provided an estimate of a can be supplied. This could be done if the data in hand provided observations near to or at the maximum size. Then for each of the size observations, y, at time t, calculate $\log_e (a/y - 1)$, and plot the results against t. If the points on the graph approximate to a straight line, then the growth data approximate to a logistic function.

The Gompertz function

In 1940, P. B. Medawar published a paper in which he deduced theoretically that the growth of the embryo chicken heart should follow a particular kind of growth function. The function concerned is known as the *Gompertz*, which was formulated as long ago as 1825 by a German mathematician of that name, in connexion with actuarial studies. The equation is

$$y = a\,e^{-be^{-kt}} \qquad (9.41)$$

and the logarithmic form is

$$\log_e y = \log_e a - b\,e^{-kt} \qquad (9.42)$$

The curves of *functions 9.41* and *9.42* are shown in Fig. 9.5, (a) and (b), respectively.

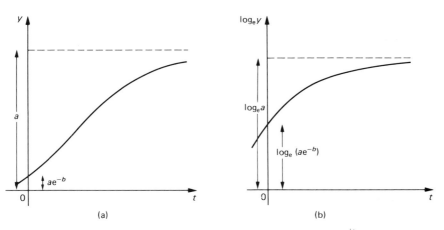

(a)   (b)

Fig. 9.5 The Gompertz function: (**a**) curve of the function $y = a\,e^{-be^{-kt}}$, and (**b**) curve of the function $\log_e y = \log_e a - b\,e^{-kt}$.

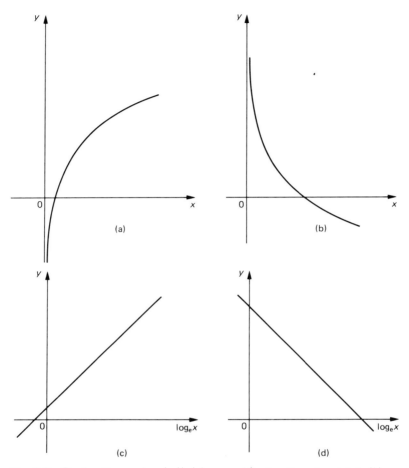

Fig. 9.2   The function $y = \log_e (ax^b)$: (**a**) curve of the function when $b > 1$, (**b**) curve of the function when $b < 1$, (**c**) the straight line $y = \log_e a + b \log_e x$ corresponding to (a), (**d**) the straight line corresponding to (b).

In this example, the portion of the curve below the x-axis is biologically meaningless, since it would be most unlikely that very low concentrations of the physiologically active substance would induce a negative response, when at higher concentrations a positive response is obtained. Also, at very high concentrations, there would be a maximum response by the organism, and this would be approached asymptotically; but the logarithmic function has no upper asymptote at any value of y. For these reasons, the logarithmic function can only provide an empirical description *within the range of the data*, that is, only over the range of concentrations of the physiologically active substance included in the experiment. Compare with the Michaelis–Menten function (pages 57–60).

Differentiating *9.41* with respect to time, using the function of a function rule, gives

$$\frac{dy}{dt} - akb\, e^{-kt}\, e^{-be^{-kt}}$$

and dividing both sides by y gives the relative growth rate as a function of time:

$$R = kb\, e^{-kt} \tag{9.43}$$

Taking logarithms of both sides of *9.43* gives

$$\log_e R = \log_e (kb) - kt \tag{9.44}$$

Thus, the logarithm of relative growth rate is a linear function of time, with gradient $-k$ and intercept $\log_e (kb)$.

Another linear relationship involving relative growth rate may be derived as follows. From *9.43*, we have

$$b\, e^{-kt} = \frac{R}{k}$$

and from *9.42* $b\, e^{-kt} = \log_e a - \log_e y$

Since the left-hand sides of the above two equations are equal, we may equate the right-hand sides:

$$\frac{R}{k} = \log_e a - \log_e y$$

i.e. $R = k \log_e a - k \log_e y \tag{9.45}$

Equation 9.45 shows that the relative growth rate is a linear function of the logarithm of size, with gradient $-k$ and intercept $k \log_e a$. Thus the Gompertz function has two linear relationships involving relative growth rate, one with time, and the other with the logarithm of size.

Finally, we may obtain a linear form of the growth function itself from *equation 9.42*:

$$\log_e \left\{ \log_e \left(\frac{a}{y} \right) \right\} = \log_e b - kt \tag{9.46}$$

Medawar's growth model

When Medawar deduced theoretically that the growth of the embryo chicken heart should follow a Gompertz function, he used the concept of 'growth energy'. He argued that this growth energy should be assessable through the application of a specific growth inhibitor; the more inhibitor required to stop growth, the higher the growth energy.

The experimental basis consisted of the application of just sufficient inhibitor to make the growth of the tissue cease, at a series of different ages. Then, assuming that growth energy was directly proportional to the amount of inhibitor required, it was found that when the logarithm of growth energy, $\log_e G$, was plotted against time, t, the trend of the points was linear with a negative gradient

$$\log_e G = \log_e A - Kt \qquad (9.47)$$

where A and K are constants. Medawar also found a linear relationship to exist between growth energy and the logarithm of the mass of the tissue, $\log_e W$,

$$G = B - K \log_e W \qquad (9.48)$$

where B and K are constants. If the forms of *9.47* and *9.48* are compared with *9.44* and *9.45*, they are found to be identical. Thus, growth energy is a measure of relative growth rate, and the growth of the tissue must conform to the Gompertz function

EXERCISES

1. Plants of birch (*Betula pubescens*) and sycamore (*Acer pseudoplatanus*) were grown from seed, and harvests were taken at two-weekly intervals during the first season. Growth was found to be exponential for 18 weeks, and the equations fitted to the data were:

for birch: $\qquad W_B = 0.0441 \; e^{0.3041t}$

for sycamore: $\quad W_S = 0.3837 \; e^{0.2243t}$

where W was measured in g, and t in weeks from the first harvest.
(a) Write down the relative growth rate of each species.
(b) Write down the weight of each species at the first harvest.
(c) Calculate the weight of each species at the end of the exponential phase.
(d) If the above growth function had been applicable over a sufficiently long period, find the time when the weight of each of the two species would be identical.

2. In a sample of NaH_2PO_4 containing a known amount of ^{32}P it was found that 95.24% of the ^{32}P remained after 24 hours. What is the half-life of ^{32}P?

3. In an experiment to determine the effects of the plant hormone abscisic acid on the growth of isolated lengths of oat coleoptiles, it was found that the following equation empirically described the data:

$$y = \log_e (354 \; 123x^{-0.4721})$$

where x is the hormone concentration in parts per million, and y is the final length of the coleoptiles in mm 4 hours after treatment with the hormone. What is the expected final length of the coleoptiles after 4 hours when treated with:
(a) 0.1 ppm, and   (b) 5 ppm of abscisic acid?

4. Differentiate the following functions with respect to x:
(a) $e^{(a+bx)}$ (b) $e^{(1-2x+x^2)}$ (c) $1 + b\,e^{-kx}$ (d) $\log_e (a + bx + cx^2)$

5. Find (a) $\int dx/(1+x)$ (b) $\int dx/(3x-7)$ (c) $\int (1 + b\,e^{-kt})\,dt$

6. A colony of *Drosophila* flies was bred on an artificial medium, and the number of individuals, n, at time t (days) was found to conform to the logistic function:

$$n = \frac{230}{1 + 6.8888\,e^{-0.1702t}}$$

(a) Write down the theoretical maximum number of individuals in the colony.
(b) How many flies were originally present in the colony?
(c) How many flies would be expected on day 10?
(d) Rearrange the equation to give t in terms of n, instead of n in terms of t. Then calculate the time by which the maximum number is effectively reached ($n = 229.5$).

10

Plant growth analysis: a biological application of calculus

Growth of micro-organisms, plants, and animals

We have already considered at some length various aspects of the mathematical analysis of the growth of a microbial culture. First, experimental data were plotted on a graph of cell number against time (Fig. 6.5a), and then we assumed that if sampling variability were absent the growth of the culture should follow some smooth curve, e.g. as shown in Fig. 6.5(b). Then, instead of merely trying a number of different mathematical functions to see if we could find a curve which would fit the data well, we tried to build a dynamic mathematical model of the growth of the culture based on known biological facts and principles. Such considerations led us to the differential equation *6.12*, which related rate of increase in cell number at any instant of time to cell number actually present at that time. Solving this differential equation gave another function relating cell number to time; this is the function we were seeking to fit to our graph of the experimental data. In Chapter 8, we saw that the solution of the differential equation *6.12* gave rise to the logistic function.

When it comes to higher organisms, with multicellular bodies, things are not so simple. One can, of course, regard the multicellular organism as a colony of individual cells, but the main difference between a colony of unicellular organisms and a single multicellular organism is that in the latter there is a considerable interaction between the constituent cells. This results in a growth promotion of some cells, and an inhibition in the growth of others. There is 'division of labour' between cells, and so they differentiate into various types of tissue. Hence it might appear that a simplified model of growth on the lines developed earlier in the book will not suffice for higher organisms.

Nevertheless, growth data obtained for such organisms (i.e. measurements of a certain aspect of size, usually fresh weight in animals and dry weight in plants) often show curves which are similar in form to that developed for a microbial colony. The main difference will be in the position of the point of inflexion; in the logistic function, this point of maximum absolute growth rate is mid-way between zero and maximum size, while in the growth curves of higher organisms the point of inflexion is usually elsewhere.

Among the higher organisms, there is a fundamental difference between animals and plants in respect of growth. The animal body normally shows a definite period of growth, the immature or juvenile phase, and when the animal is mature growth usually ceases. During the growth phase, the different parts of the animal body (head, limbs, etc.) are normally growing in a synchronized manner; in fact, it is found that the relative growth rates of the different parts of the animal's body bear a constant ratio to one another (page 51). Because of the cessation of size increase at maturity, growth is said to be **determinate**.

The situation is different in plants. In flowering plants, an annual species is a familiar and simple example. It is still true that we can distinguish a juvenile phase when vegetative growth is being made, and an adult phase when growth of the reproductive organs is occurring. The latter marks an important difference between plants and animals, in that the reproductive parts of a plant form a much greater proportion of the overall size than do the reproductive organs of animals. It is also true that the relative growth rates of the major parts of the plant – foliage, stem, roots – may bear a constant relationship to one another for prolonged periods of time, but it must be remembered that each of these main organ systems does not have a homogeneous structure. Thus the foliage is made up of leaves all of different ages, the stem is likewise composed of many internodes, and even the roots can be subdivided in a similar manner.

Many plants continue to grow vegetatively even during the reproductive phase; for example, new leaves continue to be initiated at the shoot apex if the inflorescence is axillary. Now although the growth of an *individual* leaf is determinate, that is it grows to a certain maximum size and then stops, the growth of the foliage system is **indeterminate** since the ability of the plant to produce new leaves is not usually definitely circumscribed. In fact, since each of the main organ systems on a plant is composed of sub-units, the growth of the whole plant is indeterminate. Hence, if the growth of an annual plant is followed from germination to senescence, although the growth data may tend to approach something resembling a sigmoid curve (when the growth attribute is plotted against time), the whole situation is much more nebulous than in the case of animals or micro-organisms. Even if a mathematical function could be found to fit growth data obtained from whole plants, it would not be very meaningful for two reasons. Firstly, the function would tend to be so complex that a useful biological interpretation of it would be very difficult, and secondly, the indeterminacy of growth would lead to a model which was not simple enough for the constants in it to have a useful biological meaning. Hence for plants, the analysis of growth is not attempted in terms of a fitted mathematical function (but see page 179), but methods have been developed, based directly on the actual growth data themselves, which enable useful biological information to be obtained. The scheme, which is known simply as '*plant growth analysis*', is described here as it affords a good illustration of an application of the calculus to a problem in biology.

Plant growth analysis: principles

Although the methods of plant growth analysis do not rely on any mathematical function fitted to the growth data, it is convenient in developing the ideas to imagine a group of annual plants whose growth happens to conform very closely to that of a logistic function. This means that the growth curve of the plants will be as shown in Fig. 10.1(a). There are many ways of defining and measuring plant growth, and the method selected will depend on the purpose of the study. Usually, growth studies in plants are enquiries into their productivity, measured in terms of their increase in non-aqueous material; so the attribute measured is dry mass, or as it is more commonly (but erroneously) called, dry weight. As dry weight is a destructive measurement, it is not possible to make a series of measurements on one plant; a group of similar plants must be used, and the assumption is made that the patterns of growth shown by the individual plants in the group are very similar.

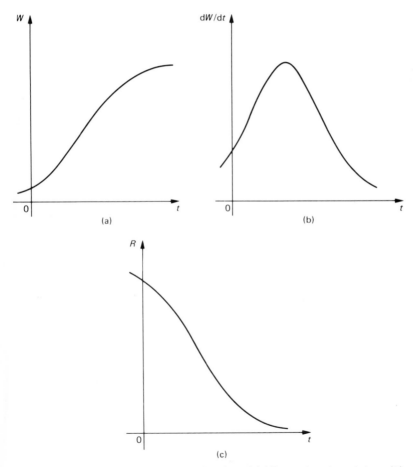

Fig. 10.1   The logistic as a growth function: (**a**) W as a function of time, (**b**) absolute growth rate dW/dt as a function of time, (**c**) relative growth rate $(1/W)$ (dW/dt) as a function of time.

Dry weight is normally given the symbol of W, while time is denoted by t as usual. With this notation, absolute growth rate is dW/dt, and relative growth rate is $(1/W)\,(dW/dt) = R$. When the $W = f(t)$ relationship is logistic, then the curves of absolute and relative growth rates with respect to time are as shown in Fig. 10.1(b and c), respectively. As remarked in the previous chapter (page 161), absolute growth rate is not a useful concept, and in fact does not appear in the methodology of plant growth analysis; thus we need concern ourselves no longer with absolute growth rate. The curve of relative growth rate declines throughout growth in a 'reverse sigmoid' manner: the change in relative growth rate as the plant grows is an important factor in plant growth analysis, and we shall return to consider the measurement of relative growth rate later on. Meanwhile, there are other quantities involved in plant growth analysis, which must now be considered.

The partitioning of relative growth rate

On page 162 it was stated that the relative growth rate of an organism is a measure of the efficiency of the material comprising the organism in the production of new material, and evidently this measure is an average over the whole organism. Now, as we have already pointed out, a higher plant is far from homogeneous, and so a measure of the average efficiency of dry matter production is limited in the information it can convey.

By far the largest part of the dry matter of a plant is in the form of organic compounds, whose primary constituent, carbon, was originally assimilated into the plant through the leaves in the form of carbon dioxide. Only a small fraction of the plant's dry mass consists of mineral elements assimilated from the soil via the roots. Hence if, as a first approximation, the mineral content is disregarded, the leaves are the sole assimilatory organs by means of which material external to the plant becomes incorporated into the plant body. Therefore, any measure of the overall efficiency of the plant depends partly on the leaves; and so the actual value of the relative growth rate at an instant of time will depend on the efficiency of the leaves in assimilation, and the amount of foliage present, at that time. If the latter is measured as the proportion of leaf dry weight to dry weight of the whole plant, we can write

$$R = E \cdot \frac{L}{W} \qquad (10.1)$$

where L is the dry weight of the foliage of the plant, and W is the dry weight of the whole plant (including the leaves). The ratio L/W evidently represents the amount of foliage relative to the whole plant, and so E is a measure of the efficiency of the leaves in the production of new dry matter. Expanding the left-hand side of *10.1*, we have

$$\frac{1}{W} \cdot \frac{dW}{dt} = E \cdot \frac{L}{W}$$

and cross-multiplication yields

$$E = \frac{1}{L} \cdot \frac{dW}{dt} \qquad (10.2)$$

The right-hand side of this expression is very similar to relative growth rate; it is the absolute growth rate of the whole plant divided by leaf dry weight, or the absolute growth rate of the plant per unit leaf dry weight, and can be interpreted as the efficiency of the leaf material in the production of new dry matter. The quantity E is called the **net assimilation rate**, or **unit leaf rate**, and may be regarded as a measure of net photosynthetic rate. The ratio L/W in *relationship 10.1* is termed the **leaf weight ratio**.

The partitioning of unit leaf rate

When measuring the photosynthetic efficiency of a plant, the plant physiologist usually expresses the rate per unit of leaf area. This attribute of the leaf is usually more relevant in photosynthetic studies than is leaf dry weight. It is the area of the laminae which determines how much radiation is intercepted, and the number of stomata, through which carbon dioxide diffuses, is more dependent on leaf area than on leaf weight. As we have defined it in the previous paragraph, it would appear that unit leaf rate must be assessed per unit of leaf dry weight. Yet this is not so. Suppose that unit leaf rate is defined per unit leaf area, then *10.2* becomes

$$E_A = \frac{1}{L_A} \cdot \frac{dW}{dt}$$

The suffix, A, indicates that we are using leaf area, L_A, and are measuring unit leaf rate on a leaf area basis, E_A. Unit leaf rate on a leaf weight basis *10.2* could be written as

$$E_w = \frac{1}{L_w} \cdot \frac{dW}{dt}$$

and it is E_w which enters into the primary partitioning of relative growth rate, *equation 10.1*. Now

$$\left.\begin{array}{c} \dfrac{1}{L_w} \cdot \dfrac{dW}{dt} = \dfrac{1}{L_A} \cdot \dfrac{dW}{dt} \times \dfrac{L_A}{L_w} \\[3mm] \text{i.e.} \quad E_w = E_A \times \dfrac{L_A}{L_w} \end{array}\right\} \qquad (10.3)$$

Substituting for E_W (i.e. E) in *10.1*, we finally have

$$R = E_A \times \frac{L_A}{L_W} \times \frac{L_W}{W} \tag{10.4}$$

where L_W has been written instead of L in the last term on the right-hand side.

Thus relative growth rate, which measures the efficiency of the plant as a whole in the production of dry matter, has been resolved into three components, each with a distinct physiological or morphological meaning. Unit leaf rate, E_A, is a measure of net photosynthetic rate, as already mentioned. Very broadly, photosynthesis may be regarded as a process with an input and an output; the input is carbon dioxide and water, and the output is oxygen plus organic compounds. Unit leaf rate is a unique method of assessing photosynthetic rate in that it does so by measuring the rate of production of organic matter. Usually methods of measuring the photosynthetic rate involve the rate of input of carbon dioxide or the rate of output of oxygen. Methods based on gaseous exchange measure the photosynthetic rate over a relatively short time period, whereas unit leaf rate is, as we shall see later, a long-term measure.

The second term on the right-hand side of *10.4* is the ratio of leaf area to leaf dry weight, and is called the **specific leaf area**. To some extent it is a measure of leaf thickness, but there is more to it than that. Consider a single leaf: as a primordium it consists of meristematic cells without vacuoles and intercellular spaces; there is plenty of cellular material but very little area, and hence the specific leaf area is low. As the leaf expands, cell division slows down, and cell extension, with its accompanying vacuolation and formation of intercellular spaces, becomes predominant. Area increases faster than dry mass, and so the specific leaf area rises. After maximal area has been reached, accumulation of dry matter still persists in the form of deposition of extra cell wall material, and the specific leaf area declines. Specific leaf area may thus be regarded as a measure of leaf density.

The final term, L_W/W, is the leaf weight ratio, and is quite simply the ratio of leaf dry weight to total plant dry weight. Hence, this ratio lies in the range $0 \leqslant L_W/W < 1$. In the very young seedling of species whose seeds contain endosperm, the leaf weight ratio is small at first, while the seedling is dependent on endosperm reserves. As the plant becomes nutritionally independent, the leaf weight ratio increases rapidly and may reach a value of about 0.8 in herbaceous species, but rather lower in woody types. In many herbaceous plants the leaf weight ratio stays constant during vegetative growth, especially if the form of growth is a rosette. When the stem grows, the leaf weight ratio falls; again, the decline in leaf weight ratio is particularly marked in woody species. During the reproductive phase, the decline in leaf weight ratio is accelerated.

Calculation of the quantities involved in growth analysis

Specific leaf area and leaf weight ratio

These ratios present no problem. The primary data for a growth analysis consist of determinations of the dry weights of the leaves and the rest of the plant, together with an estimate of total leaf area, for each member of a batch of plants taken at a particular point in time. The batch of plants taken at a particular time is known as a harvest; several harvests are taken, usually at regular intervals. At each of the harvests, specific leaf area and leaf weight ratio are simply evaluated from the data.

Relative growth rate

This parameter has been defined as $(1/W)\,(dW/dt)$; it contains a differential coefficient, which implies that it is defined at an instant of time. It is not possible to measure an instantaneous rate of change, and even if it were, the result would not be meaningful. To see this, remember that plant growth has been defined in terms of change in dry weight, which is a direct consequence of photosynthesis and respiration. Now the primary environmental factor affecting photosynthetic rate, and hence the rate of dry weight increase, is light intensity. Photosynthetic rate in this context is net, i.e. photosynthetic rate minus respiration rate, and evidently it will be negative at night. Therefore, relative growth rate will also be negative at night. Moreover, on partly cloudy days the relative growth rate will be frequently changing. Even without short term changes in light intensity during daylight hours due to cloud, a curve of relative growth rate against time would look like that in Fig. 10.2 over a

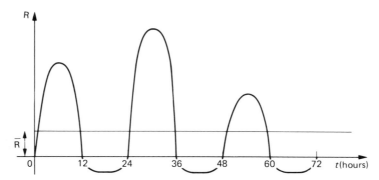

Fig. 10.2   A typical curve of relative growth rate against time for a plant over a three-day period. The day length is 12 hours and it is assumed that there are no short-term fluctuations in light intensity. The mean relative growth rate, R, for the 72 hour period is also shown.

succession of days and nights. Hence at any instant of time, a value for relative growth rate is unique.

However, it is readily apparent that over the three-day period shown in Fig. 10.2, a *mean* relative growth rate, $\bar{R}$, is a valid concept (page 123), and it is shown as a horizontal line on the graph. So if we imagine that a harvest is taken at sunrise on the first day (at $t = 0$), and another harvest is taken at sunrise on the fourth day, the interval between harvests is 72 hours. How can we calculate the mean relative growth rate over the three-day period?

The principle of the method for answering this question was given in Chapter 7 (page 124). The curve shown in Fig. 10.2 must be the curve of some function relating relative growth rate to time, i.e. a function of the form

$$R = f(t) \tag{10.5}$$

We may never know what $f(t)$ is; indeed it is highly unlikely that we could ever find a function which would generate the curve in Fig. 10.2 exactly. However, for the situation shown, and measuring time in days, we have from *equation 7.29* that

$$\bar{R} = \frac{1}{3 - 0} \int_0^3 f(t)\, dt$$

More generally, we consider two harvests, one at time t_1 and the other at time t_2; when the mean relative growth rate between these times is given by

$$\bar{R} = \frac{1}{t_2 - t_1} \int_{t_1}^{t_2} f(t)\, dt \tag{10.6}$$

Now although the form of $f(t)$ is unknown, it can be replaced by R, see *10.5*, and in turn, R can be replaced by $(1/W)(dW/dt)$. Thus *10.6* becomes

$$\bar{R} = \frac{1}{t_2 - t_1} \int_{t_1}^{t_2} \frac{1}{W} \cdot \frac{dW}{dt} \cdot dt \tag{10.7}$$

Now in *10.7* to the right of the integral sign, the dt's cancel leaving dW/W. So the variable of integration is now dW instead of dt, and because of this the limits of integration cannot remain as t_1 and t_2. Let plant weight at the first harvest be W_1 and at the second harvest W_2, so *10.7* becomes

$$\bar{R} = \frac{1}{t_2 - t_1} \int_{W_1}^{W_2} \frac{dW}{W} \tag{10.8}$$

Note that the $1/(t_2 - t_1)$ outside the integral sign remains in terms of t because it is a constant. Integrating

$$\bar{R} = \frac{1}{t_2 - t_1} \left[\log_e W \right]_{W_1}^{W_2}$$

(see standard *integral 9.24*), and so

$$\bar{R} = \frac{\log_e W_2 - \log_e W_1}{t_2 - t_1} \tag{10.9}$$

Since the original function relating R and t, *10.5*, is completely unknown, and no assumptions have been made about it in the above derivation, *equation 10.9* will always give the correct mean relative growth rate between two harvests regardless of the way the plants are growing during the interval.

In this chapter, two graphs depicting curves of R against time have been presented (Fig. 10.1c and Fig. 10.2) which look very different from one another. The true situation for growing plants is shown in Fig. 10.2, while the curve shown in Fig. 10.1(c) has been derived from a relatively simple mathematical function relating increase in plant weight to time. This latter function is continually increasing, and so can never yield a negative relative growth rate which undoubtedly occurs at night. So the curve shown in Fig. 10.1(c) should be regarded more as a curve of mean relative growth rate against time: it is a smooth approximation to an $\bar{R}-t$ curve rather than an $R-t$ curve.

Unit leaf rate

All the remarks made about relative growth rate in the first two paragraphs of the previous section apply equally to unit leaf rate, and so in practice we calculate the mean unit leaf rate between two harvests. Suppose that unit leaf rate is related to time by some function

$$E = \phi(t) \tag{10.10}$$

Then between two harvests at t_1 and t_2 the mean unit leaf rate is given by

$$\bar{E} = \frac{1}{t_2 - t_1} \int_{t_1}^{t_2} \phi(t) \, dt$$

Now $\phi(t) = (1/L)(dW/dt)$, where L is leaf area in this context, and so

$$\bar{E} = \frac{1}{t_2 - t_1} \int_{t_1}^{t_2} \frac{1}{L} \cdot \frac{dW}{dt} \, dt \tag{10.11}$$

As in the similar expression for relative growth rate *10.7*, the d*t*'s cancel leaving d*W* as the variable of integration, and so the limits t_1 and t_2 are replaced by W_1 and W_2:

$$\bar{E} = \frac{1}{t_2 - t_1} \int_{W_1}^{W_2} \frac{dW}{L} \qquad (10.12)$$

The integral in *10.12* cannot be evaluated, however, unless the relationship between W and L is known or assumed.

We shall briefly discuss relationships between parts of organisms later in this chapter, but usually, however, during the vegetative phase of plant growth it is safe to assume that there is a linear relationship between total plant weight and leaf area, i.e.

$$W = a + bL \qquad (10.13)$$

and the error involved if the relationship departs appreciably from linearity is minimized if the harvest is such that leaf area no more than doubles from one harvest to another. Differentiating *10.13* gives

$$\frac{dW}{dL} = b$$

or $\quad dW = b\,dL \qquad (10.14)$

On using *10.14* to substitute for dW in *10.12*, we again have a new variable of integration, dL, and so the limits of integration in *10.12* must be changed once more. Assume that at t_1, when total plant weight is W_1, the leaf area is L_1; similarly at t_2, when total plant weight is W_2, the leaf area is L_2. Putting all these changes into *10.12*, it becomes

$$\bar{E} = \frac{1}{t_2 - t_1} \int_{L_1}^{L_2} \frac{b\,dL}{L}$$

or $\quad \bar{E} = \frac{b}{t_2 - t_1} \int_{L_1}^{L_2} \frac{dL}{L} \qquad (10.15)$

Integrating $\quad \bar{E} = \frac{b}{t_2 - t_1} \left[\log_e L \right]_{L_1}^{L_2}$

and so $\quad \bar{E} = \frac{b(\log_e L_2 - \log_e L_1)}{t_2 - t_1} \qquad (10.16)$

Now although to get this far we have had to assume a linear relationship between W and L, we do not wish to assign actual values to the constants in this relationship; fortunately this is not necessary. At the harvest at t_1 we have, from *equation 10.13* and the paragraph beneath *10.14*,

$$W_1 = a + bL_1 \qquad (10.17)$$

and at t_2
$$W_2 = a + bL_2 \qquad (10.18)$$

Subtracting *10.17* from *10.18*, we have

$$W_2 - W_1 = (a - a) + (bL_2 - bL_1)$$

i.e.
$$W_2 - W_1 = b(L_2 - L_1)$$

which gives
$$b = \frac{W_2 - W_1}{L_2 - L_1} \qquad (10.19)$$

Substituting for b in *10.16* gives

$$\bar{E} = \frac{(W_2 - W_1)(\log_e L_2 - \log_e L_1)}{(L_2 - L_1)(t_2 - t_1)} \qquad (10.20)$$

Equation 10.20 gives the correct value for the mean unit leaf rate between the two harvests regardless of how the whole plant and leaf area are growing with respect to time, but subject to the constraint that there is a linear relationship between whole plant dry weight and leaf area.

Thus all the quantities involved in plant growth analysis can be calculated from primary data obtained at each of a number of harvests – a minimum of two is required. The two ratios – leaf weight ratio and specific leaf area – are calculated from leaf weight, whole plant weight, and leaf area at each harvest. The mean relative growth rate between an adjacent pair of harvests is computed from the whole plant weights at each of the two harvests, and mean net assimilation rate between a pair of harvests by using the leaf areas and total plant weights at each of the two harvests.

Curve fitting

Although near the beginning of this chapter (page 170) we emphasised that the concepts and methods of plant growth analysis were not originally based on specific mathematical curves fitted to primary growth data, nevertheless there has been a considerable trend in this direction during recent years. This topic will not be developed here; we mention it because of its modern relevance. A full account is given by R. Hunt in his book *Plant Growth Curves*, published by Edward Arnold.

Allometry and growth analysis

If two parts of a plant, say foliage, x, and stem, y, are related allometrically as they grow (Chapter 4, page 51), then this indicates an important physiological phenomenon.

$$\text{Let} \qquad y = ax^b \qquad\qquad (10.21)$$

$$\text{then} \qquad \frac{dy}{dx} = bax^{(b-1)} \qquad\qquad (10.22)$$

Now divide both numerator and denominator of the left-hand side of *10.22* by dt

$$\frac{dy}{dt} \bigg/ \frac{dx}{dt} = bax^{(b-1)}$$

The right-hand side of *10.22* is unaffected by this procedure. Now multiply both sides by x/y, then

$$\frac{x}{y}\left(\frac{dy}{dt}\bigg/\frac{dx}{dt}\right) = \frac{baxx^{(b-1)}}{y}$$

The product $xx^{(b-1)} = x^b$; also substitute for y using *10.21*:

$$\frac{x}{y}\left(\frac{dy}{dt}\bigg/\frac{dx}{dt}\right) = \frac{bax^b}{ax^b} = b \qquad\qquad (10.23)$$

The left-hand side of *10.23* can be written in the form

$$\frac{1}{y}\cdot\frac{dy}{dt}\bigg/\frac{1}{x}\cdot\frac{dx}{dt}$$

which is the ratio of the relative growth rate of part y to the relative growth rate of part x. In keeping with the notation of *equation 9.34* (page 162), we can write the above ratio as R_y/R_x, and so *equation 10.23* finally becomes

$$\frac{R_y}{R_x} = b \qquad\qquad (10.24)$$

Thus, the gradient of the allometric line is equal to the ratio of the relative growth rates of the two plant parts. As growth in this context is increase in dry material, which is primarily derived from the carbon fixed in photosynthesis, then the gradient of the allometric line, b, quantifies the partitioning of dry matter between the two parts of the plant. Moreover, allometric growth shows that the partitioning ratio is constant over the time interval in which a single allometric function can be said to describe the growth data. Indeed, as already shown (Fig. 4.5), over an extended duration more than one allometric function may be needed to describe the data and involving a sharp change of gradient – implying a sharp change in the pattern of assimilate partitioning.

EXERCISE

The following data are the leaf areas, L_A, leaf dry weights, L_w, and total plant dry weights, W, of two series of plants of *Browallia speciosa* at a series of times, t. One series was grown in full daylight, and the other was grown under a screen admitting about 40% of full daylight. Each figure is the mean of 4 plants.

	Unshaded			Shaded		
t (days)	L_{A_2} (cm^2)	L_W (g)	W (g)	L_{A_2} (cm^2)	L_W (g)	W (g)
0	0.3000	0.0009	0.0012	0.5093	0.0007	0.0011
3	0.5000	0.0014	0.0020	0.7596	0.0011	0.0015
7	1.4787	0.0037	0.0048	0.8293	0.0011	0.0015
10	1.3835	0.0031	0.0042	1.1960	0.0019	0.0026
14	2.4177	0.0070	0.0096	2.3262	0.0034	0.0048
17	3.4185	0.0104	0.0149	2.5225	0.0044	0.0064
21	7.9747	0.0205	0.0301	2.9430	0.0043	0.0060
24	5.8193	0.0157	0.0234	5.0663	0.0074	0.0106
28	10.4533	0.0315	0.0460	6.0600	0.0100	0.0139

Calculate, for each series of plants, L_A/L_w and L_w/W at each harvest, and $\bar{R}$ and E for each harvest interval. Draw graphs of L_A/L_w, L_w/W, R, and $\bar{E}$ against t to show the effect of shading on these quantities. Try to interpret the results biologically.

11

Matrix algebra

Matrices and vectors

A *matrix*, plural *matrices*, is a collection of numbers (some or all of which may be algebraic symbols) arranged in a rectangular form with rows and columns, of which there may be any number. The whole matrix can be given a symbol of its own;

$$\text{e.g.} \qquad \mathbf{M} = \begin{bmatrix} 1 & -2 \\ 0 & 1 \end{bmatrix} \qquad (11.1)$$

In *relationship 11.1* we have a matrix consisting of two **rows** and two **columns** of figures and, to show that the four numbers must be regarded as an entity, they are enclosed in square brackets. The whole collection of numbers – the matrix – is given the symbol **M**, and matrices will always be denoted by symbols in heavy type. Hence **M** is distinguishable from M, the latter denoting a single number. The convention of enclosing the numbers within a matrix by square brackets is not universal; the use of rounded brackets is almost as common:

$$\mathbf{M} = \begin{pmatrix} 1 & -2 \\ 0 & 1 \end{pmatrix}$$

More unfortunate, however, is the fact that some authors do not use heavy type for symbolizing matrices; it is far more satisfactory to do so as ambiguities are thereby avoided.

The general matrix

The individual numbers within a matrix are called *elements*, and they are arranged in two dimensions – rows and columns. The *general matrix*, of which any particular matrix is a special case, has m rows and n columns. The usual way of writing the general matrix is to symbolize it by a capital letter in heavy type, in the usual way, and then to denote the individual elements by the corresponding ordinary lower case letter with two subscripts: the first subscript denotes the row to which the element belongs, and the second subscript the column. The general matrix can then be written out as shown below.

$$\mathbf{A} = \begin{bmatrix} a_{11} & a_{12} & a_{13} \cdots a_{1j} \cdots a_{1n} \\ a_{21} & a_{22} & a_{23} \cdots a_{2j} \cdots a_{2n} \\ a_{31} & a_{32} & a_{33} \cdots a_{3j} \cdots a_{3n} \\ \vdots & \vdots & \vdots & \vdots & \vdots \\ a_{i1} & a_{i2} & a_{i3} \cdots a_{ij} \cdots a_{in} \\ \vdots & \vdots & \vdots & \vdots & \vdots \\ a_{m1} & a_{m2} & a_{m3} \cdots a_{mj} \cdots a_{mn} \end{bmatrix} \qquad (11.2)$$

Any particular element may be designated by its row number and its column number, that is by its subscript. For example, the element of $\mathbf{A}$, above, in the 4th row and the 2nd column is a_{42}, in the 3rd row and the 3rd column is a_{33}, and more generally, in the ith row and the jth column is a_{ij}. Notice that the row specification is *always* given first, and that the subscripts are not separated from one another by commas.

   Matrix $\mathbf{A}$ in *11.2* has m rows and n columns; thus it is spoken of as an $(m \times n)$ matrix, and this is a convenient way of specifying the size of a matrix. The element a_{ij} is called the **general element.**

The transpose of a matrix

The **transpose** of a matrix is simply the matrix with its rows and columns interchanged, and is denoted by priming the capital letter designating the matrix. Thus, the transpose of $\mathbf{M}$ of *11.1* is given by

$$\mathbf{M}' = \begin{bmatrix} 1 & 0 \\ -2 & 1 \end{bmatrix}$$

The symbol $\mathbf{M}^{\mathrm{T}}$ is also used to denote a transposed matrix.

Particular types of matrix

Having introduced some points of terminology and notation, and also the general matrix as a framework of reference, we are now in a position to examine some important specific kinds of matrix.

Zero matrices

A *zero matrix* is one in which all elements are zero. It is denoted by **O**, and an example is

$$\mathbf{O} = \begin{bmatrix} 0 & 0 & 0 & 0 \\ 0 & 0 & 0 & 0 \\ 0 & 0 & 0 & 0 \end{bmatrix}$$

Square matrices

A *square matrix* is one in which the numbers of rows and columns are equal, and such matrices are of outstanding importance. The size can be indicated either by specifying that the matrix is an $(m \times m)$ one, or that the matrix is of *order* m. Matrix **M** of *11.1* is a square matrix of order 2. Matrix **A** of *11.3* below is a (4×4) matrix, or alternatively a square matrix of order 4.

$$\mathbf{A} = \begin{bmatrix} 2 & -1 & -4 & 5 \\ 2 & 1 & -3 & 2 \\ 2 & 3 & 0 & 2 \\ 2 & 5 & 5 & 7 \end{bmatrix} \tag{11.3}$$

There are two categories of *diagonal elements* in a square matrix; one from top-left to bottom-right, and the other from top-right to bottom-left. The former diagonal is particularly important, and is known as the *leading diagonal*; while the diagonal from top-right to bottom-left has no significance at all.

Symmetric matrices

A *symmetric matrix* is a square matrix which is identical with its transpose. Thus

$$\mathbf{S} = \begin{bmatrix} 3 & -1 & 2 \\ -1 & 0 & 1 \\ 2 & 1 & 4 \end{bmatrix} \tag{11.4}$$

is a symmetric matrix, as it is unchanced when rows and columns are interchanged. The name arises because the values of the elements are symmetrically disposed about the leading diagonal; it can be said that the elements on one side of the leading diagonal are 'reflected' on the other side, and the leading diagonal acts as though it were a plane mirror. Because of this, a symmetric matrix need not be quoted in full, but in one of two partial ways.

Thus matrix **S** of *11.4* can be written either as

$$\begin{bmatrix} 3 & -1 & 2 \\ & 0 & 1 \\ & & 4 \end{bmatrix} \quad \text{or as} \quad \begin{bmatrix} 3 & & \\ -1 & 0 & \\ 2 & 1 & 4 \end{bmatrix}$$

Diagonal matrices

A *diagonal matrix* is a square matrix in which all elements not on the leading diagonal are zero.

$$\text{Thus} \quad \mathbf{D} = \begin{bmatrix} 2 & 0 & 0 \\ 0 & -1 & 0 \\ 0 & 0 & -5 \end{bmatrix} \qquad (11.5)$$

is a diagonal matrix. Such matrices are particularly simple to deal with.

Unit matrices

The *unit matrix* is a diagonal matrix in which all elements of the leading diagonal are unity. The unit matrix is always denoted by **I** or as $\mathbf{I}_n$, where n is the order of the matrix; for example

$$\mathbf{I}_4 = \begin{bmatrix} 1 & 0 & 0 & 0 \\ 0 & 1 & 0 & 0 \\ 0 & 0 & 1 & 0 \\ 0 & 0 & 0 & 1 \end{bmatrix} \qquad (11.6)$$

is the 4th order unit matrix. Unit matrices have interesting properties, and play a leading role in matrix algebra.

Column vectors

If in the general matrix $n = 1$, then the resulting matrix has only one column with m rows. A logical name would be 'column matrix', but it is more usually known as a *column vector* for a reason that will be explained later (page 186).

A column vector is normally denoted by a small (lower case) letter in heavy type, e.g.

$$\mathbf{a} = \begin{bmatrix} a_{11} \\ a_{21} \\ a_{31} \\ \vdots \\ a_{i1} \\ \vdots \\ a_{m1} \end{bmatrix} \qquad (11.7)$$

Row vectors

Referring again to the general matrix, if $m = 1$ then the resulting matrix has only one row with n columns, and is known as a *row vector*. In theory and in applications, the column vector is the more fundamental, and so a row vector is usually regarded as the transpose of a column vector and is denoted by a primed small letter in heavy type:

$$\mathbf{a}' = [a_{11} \quad a_{12} \quad a_{13} \ldots a_{1j} \ldots a_{1n}] \qquad (11.8)$$

The constant subscript 1 may be omitted in a row vector, the elements being labelled a_1 to a_n; and similarly in a column vector the elements will be labelled a_1 to a_m.

Geometrical interpretation of a column vector

If you have pursued a course in mechanics or elementary physics, you may remember that a **vector** is a quantity having both magnitude and direction, as opposed to a **scalar** quantity which has magnitude only. Thus, ordinary numbers are scalar quantities, but a force, which has a magnitude and is applied in a particular direction, is a vector quantity. Row and column vectors are not, however, called vectors because they are orientated horizontally and vertically, respectively, but for an entirely different reason.

Consider the column vector $\mathbf{x} = \begin{bmatrix} 2 \\ 1 \end{bmatrix}$. Now the two numbers within the column vector can be thought of as the co-ordinates of a point P situated in a plane defined by the usual x–y axes (Fig. 11.1). But the column vector is not merely another way of specifying the co-ordinates of a point; it means something more than that. Join the point, whose co-ordinates are given by the elements of the column vector, to the origin by a straight line. This line is a vector quantity; it has both magnitude (i.e. length), given by

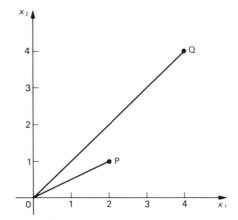

Fig. 11.1   Vectors OP and OQ (see page 186).

$\sqrt{\{(2)^2 + (1)^2\}} = \sqrt{5}$ (see page 24), and direction or orientation, given by $\tan^{-1}(1/2) \simeq 26°34'$ (i.e. the angle whose tangent is $\frac{1}{2}$). The angle is always measured in an anti-clockwise direction from the horizontal axis. Similarly, another vector $\mathbf{y} = \begin{bmatrix} 4 \\ 4 \end{bmatrix}$ represents the straight line OQ with a length of $\sqrt{\{(4)^2 + (4)^2\}} = \sqrt{32}$, and an orientation of $\tan^{-1}(4/4) = 45°$.

Thus we see that a two-component column vector can represent a straight line in two-dimensional space from the origin to the point whose co-ordinates are specified by the elements of the column vector. By analogy, a three-component column vector may represent a straight line in three-dimensional space, again from the origin to the point whose components are specified by the elements of the vector. The analogy may be pursued further, and we conclude that an *n*-component column vector can represent a similar straight line in a space of *n* dimensions; it cannot be illustrated, but it can be described mathematically.

Elements of matrix algebra

Since any kind of matrix can be given a unique symbol, this means that the symbol represents the whole array of numbers within the matrix. If we have two different matrices, each represented by a single symbol, one might be able to operate on these symbols as can be done with symbols representing single numbers in ordinary algebra. For instance, let **A** and **B** denote two matrices; then one *could* write such expressions as **A** + **B**, and **AB**. If such operations could be performed, then presumably the results would be other matrices, i.e.

$$\left.\begin{array}{c} \mathbf{A} + \mathbf{B} = \mathbf{C} \\ \text{and} \quad \mathbf{AB} = \mathbf{D} \end{array}\right\} \qquad (11.9)$$

Hence, three questions may be posed. (1) Are operations such as those represented by the equations in *11.9* possible? (2) If so, under what conditions are they possible? (3) What are the new matrices, e.g. **C** and **D**, which are formed by such operations? The answer to Question 1 is mostly affirmative, and in considering the various facets of matrix algebra, the answers to Questions 2 and 3 will emerge.

Matrix equality

Two matrices are equal to one another only if each contains the same number of rows and the same number of columns, and if *all* corresponding elements are equal to one another. For the following two matrices, **A** and **B**, we have

$$\begin{bmatrix} a_{11} & a_{12} & a_{13} \\ a_{21} & a_{22} & a_{23} \end{bmatrix} = \begin{bmatrix} b_{11} & b_{12} & b_{13} \\ b_{21} & b_{22} & b_{23} \end{bmatrix} \qquad (11.10)$$
$$\mathbf{A} = \mathbf{B}$$

only if $a_{11} = b_{11}$, $a_{21} = b_{21}$, ..., $a_{23} = b_{23}$. To put the condition more briefly, we write

$$a_{ij} = b_{ij} \qquad \text{for all} \qquad i, j$$

Addition and subtraction

The operations of addition and subtraction can be performed and, being very similar, they can be considered together. Two matrices can be added to, or subtracted from, one another only if each contains the same number of rows and the same number of columns; and the operations are performed simply by adding or subtracting corresponding elements.

Example 11.1

$$\text{If} \qquad \mathbf{A} = \begin{bmatrix} 4 & 6 \\ 5 & 7 \end{bmatrix} \qquad \text{and} \qquad \mathbf{B} = \begin{bmatrix} 2 & 3 \\ 1 & 4 \end{bmatrix}$$

Find
(a) $\mathbf{A} + \mathbf{B}$ (b) $\mathbf{B} + \mathbf{A}$ (c) $\mathbf{A} - \mathbf{B}$ (d) $\mathbf{B} - \mathbf{A}$

(a)
$$\begin{bmatrix} 4 & 6 \\ 5 & 7 \end{bmatrix} + \begin{bmatrix} 2 & 3 \\ 1 & 4 \end{bmatrix} = \begin{bmatrix} (4+2) & (6+3) \\ (5+1) & (7+4) \end{bmatrix} = \begin{bmatrix} 6 & 9 \\ 6 & 11 \end{bmatrix}$$

(b) Proceeding in the same way as above:

$$\begin{bmatrix} 2 & 3 \\ 1 & 4 \end{bmatrix} + \begin{bmatrix} 4 & 6 \\ 5 & 7 \end{bmatrix} = \begin{bmatrix} 6 & 9 \\ 6 & 11 \end{bmatrix}$$

(c) $\begin{bmatrix} 4 & 6 \\ 5 & 7 \end{bmatrix} - \begin{bmatrix} 2 & 3 \\ 1 & 4 \end{bmatrix} = \begin{bmatrix} (4-2) & (6-3) \\ (5-1) & (7-4) \end{bmatrix} = \begin{bmatrix} 2 & 3 \\ 4 & 3 \end{bmatrix}$

(d) $\begin{bmatrix} 2 & 3 \\ 1 & 4 \end{bmatrix} - \begin{bmatrix} 4 & 6 \\ 5 & 7 \end{bmatrix} = \begin{bmatrix} (2-4) & (3-6) \\ (1-5) & (4-7) \end{bmatrix} = \begin{bmatrix} -2 & -3 \\ -4 & -3 \end{bmatrix} = -\begin{bmatrix} 2 & 3 \\ 4 & 3 \end{bmatrix}$

This example shows that:

(i) $\qquad\qquad\qquad \mathbf{A} + \mathbf{B} = \mathbf{B} + \mathbf{A}$
(ii) $\qquad\qquad\qquad \mathbf{A} - \mathbf{B} = -(\mathbf{B} - \mathbf{A})$

that is, addition and subtraction of matrices are commutative processes, as is the case for single quantities in ordinary algebra. In other words, whenever two matrices are added together, the sum is the same regardless of which matrix appears first; similarly, if one matrix is subtracted from another, the same *numerical* result is obtained whichever way round the subtraction is performed, but the sign of the result depends on the order of the two given matrices.

Care must be taken when the elements in the matrices differ in sign, as in the next example.

Example 11.2

If $\mathbf{P} = \begin{bmatrix} 1 & 2 & -1 \\ -3 & -2 & 1 \\ 2 & 3 & 2 \end{bmatrix}$ and $\mathbf{Q} = \begin{bmatrix} 3 & 2 & -1 \\ 4 & 2 & -2 \\ -1 & 3 & 1 \end{bmatrix}$

find (a) $\mathbf{P} + \mathbf{Q}$ (b) $\mathbf{P} - \mathbf{Q}$

(a) $\begin{bmatrix} 1 & 2 & -1 \\ -3 & -2 & 1 \\ 2 & 3 & 2 \end{bmatrix} + \begin{bmatrix} 3 & 2 & -1 \\ 4 & 2 & -2 \\ -1 & 3 & 1 \end{bmatrix}$

$= \begin{bmatrix} \{1+3\} & \{2+2\} & \{(-1)+(-1)\} \\ \{(-3)+4\} & \{(-2)+2\} & \{1+(-2)\} \\ \{2+(-1)\} & \{3+3\} & \{2+1\} \end{bmatrix}$

$= \begin{bmatrix} 4 & 4 & -2 \\ 1 & 0 & -1 \\ 1 & 6 & 3 \end{bmatrix}$

(b)
$$\begin{bmatrix} 1 & 2 & -1 \\ -3 & -2 & 1 \\ 2 & 3 & 2 \end{bmatrix} - \begin{bmatrix} 3 & 2 & -1 \\ 4 & 2 & -2 \\ -1 & 3 & 1 \end{bmatrix}$$

$$= \begin{bmatrix} \{(1-3)\} & \{2-2\} & \{(-1)-(-1)\} \\ \{(-3)-4\} & \{(-2)-2\} & \{1-(-2)\} \\ \{2-(-1)\} & \{3-3\} & \{2-1\} \end{bmatrix}$$

$$= \begin{bmatrix} -2 & 0 & 0 \\ -7 & -4 & 3 \\ 3 & 0 & 1 \end{bmatrix}$$

The operation of addition (or subtraction) may be summarized for two general matrices. Let

$$A = \begin{bmatrix} a_{11} & a_{12} & \cdots & a_{1n} \\ a_{21} & a_{22} & \cdots & a_{2n} \\ \vdots & \vdots & & \vdots \\ a_{m1} & a_{m2} & \cdots & a_{mn} \end{bmatrix} \quad \text{and} \quad B = \begin{bmatrix} b_{11} & b_{12} & \cdots & b_{1n} \\ b_{21} & b_{22} & \cdots & b_{2n} \\ \vdots & \vdots & & \vdots \\ b_{m1} & b_{m2} & \cdots & b_{mr} \end{bmatrix}$$

Define a matrix C such that $A + B = C$, then

$$C = \begin{bmatrix} (a_{11}+b_{11}) & (a_{12}+b_{12}) & \cdots & (a_{1n}+b_{1n}) \\ (a_{21}+b_{21}) & (a_{22}+b_{22}) & \cdots & (a_{2n}+b_{2n}) \\ \vdots & \vdots & & \vdots \\ (a_{m1}+b_{m1}) & (a_{m2}+b_{m2}) & \cdots & (a_{mn}+b_{mn}) \end{bmatrix} \quad (11.11)$$

Subtraction proceeds in an identical way, with the plus signs replaced by minus signs.

Multiplication

Multiplication of a matrix by a scalar

The word 'scalar' here refers to a single number which is not an element of a matrix. It is possible to multiply a matrix by a scalar quantity. Suppose that matrix A is given by

$$A = \begin{bmatrix} a_{11} & a_{12} \\ a_{21} & a_{22} \end{bmatrix}$$

then if k is a scalar number, we have

$$k\mathbf{A} = k \begin{bmatrix} a_{11} & a_{12} \\ a_{21} & a_{22} \end{bmatrix} = \begin{bmatrix} ka_{11} & ka_{12} \\ ka_{21} & ka_{22} \end{bmatrix} \qquad (11.12)$$

So, multiplying a matrix by a scalar implies that every element in the matrix is multiplied by the scalar.

Multiplication of two matrices

The multiplication of two matrices is a feasible operation, but note at once that it is not done in a similar way to addition and subtraction; that is, one does not merely multiply together corresponding elements in the two matrices.

You will perhaps conclude from this that matrix algebra is an illogical subject. Before considering the details of matrix multiplication, let us look a little further into the nature of ordinary algebra performed with single quantities, and make some comparisons with matrix algebra.

Ordinary algebra is a 'natural' process in that its rules derive ultimately from the fundamental process of addition. Two added to three must equal five, thus giving the inescapable result that $3 + 2 = 5$. Subtraction arises naturally from addition by invoking the idea of a negative number and using it in the process of addition in the usual way; thus $3 + (-2) = 3 - 2 = 1$. Multiplication is continued addition. Thus, 3×2 is short for $3 + 3$, or $2 + 2 + 2$, which equals 6; and 3×5 is an abbreviation for $3 + 3 + 3 + 3 + 3$, or $5 + 5 + 5$, which is 15. Division is continued subtraction. To see this, consider $24 \div 8 = 3$. Starting with 24, subtract 8 leaving 16; subtract 8 again leaving 8, and subtract 8 once more which yields 0. Three subtractions of 8 from 24 give zero, i.e. $24 - 8 - 8 - 8 = 0$ or $24 - (8 + 8 + 8) = 0$. This is the same as writing $24 - (3 \times 8) = 0$. If the divisor does not go into the dividend exactly, the actual demonstration that division is continued subtraction is not quite as straightforward, but the same principle applies. Hence, as division is continued subtraction, and subtraction is merely addition involving negative numbers, we see that the four basic rules of ordinary algebra are derived from the fundamental process of addition.

Since a matrix is an entity consisting of more than one number, there is clearly no fundamental or even single way in which one matrix can be combined with another in an algebraic operation. Hence the rules in matrix algebra are not natural ones, but are *conventions*, and they have been formulated in the way they have in order to make the fullest use of being able to apply algebraic processes to whole arrays. At first sight, the condition and rule for matrix multiplication may seem very complicated and highly artificial, but practice will show that the rule is not difficult to master, and will also demonstrate the great utility of the procedure.

First of all, two matrices can be multiplied together only if the number of columns in the first matrix is equal to the number of rows in the second. For example, let

$$\mathbf{A} = \begin{bmatrix} 1 & -2 \\ 3 & 2 \end{bmatrix} \qquad \mathbf{B} = \begin{bmatrix} 4 & 0 & 2 \\ -1 & 6 & 3 \end{bmatrix}$$

Then the multiplication $\mathbf{AB}$ is possible, as the number of columns in $\mathbf{A}$ ($=2$) is equal to the number of rows in $\mathbf{B}$ ($=2$). However, there can be no product $\mathbf{BA}$ because the number of columns in $\mathbf{B}$ ($=3$) is not equal to the number of rows in $\mathbf{A}$ ($=2$). So although a given pair of matrices can be multiplied together one way round, the operation cannot necessarily be performed with the order of the matrices reversed. This situation, which is a consequence of the rule for the multiplication of matrices, is clearly very different from the situation involving single quantities in ordinary algebra. In the latter case, for two numbers a and b, $ab = ba$; multiplication is commutative. However, for two matrices, $\mathbf{A}$ and $\mathbf{B}$, although $\mathbf{AB}$ may be feasible, $\mathbf{BA}$ may be impossible. Thus, multiplication of matrices is not, in general, commutative. Even though in a particular instance both the operations $\mathbf{AB}$ and $\mathbf{BA}$ are possible, the resultant matrices will usually be different (*example 11.4a* and b).

Returning to the two matrices introduced above, let us form the matrix product $\mathbf{AB}$. The rule states that we must combine the rows of the first matrix with the columns of the second, in the following way:

$$\begin{bmatrix} 1 & -2 \\ 3 & 2 \end{bmatrix} \begin{bmatrix} 4 & 0 & 2 \\ -1 & 6 & 3 \end{bmatrix}$$

$$= \begin{bmatrix} \{(1)(4) + (-2)(-1)\} & \{(1)(0) + (-2)(6)\} & \{(1)(2) + (-2)(3)\} \\ \{(3)(4) + (2)(-1)\} & \{(3)(0) + (2)(6)\} & \{(3)(2) + (2)(3)\} \end{bmatrix}$$

$$= \begin{bmatrix} \{4 + 2\} & \{0 + (-12)\} & \{2 + (-6)\} \\ \{12 + (-2)\} & \{0 + 12\} & \{6 + 6\} \end{bmatrix} = \begin{bmatrix} 6 & -12 & -4 \\ 10 & 12 & 12 \end{bmatrix}$$

Notice that we have multiplied a (2×2) matrix by a (2×3) one, and obtained a (2×3) product matrix: we could write

$$(2 \times 2)(2 \times 3) = (2 \times 3) \qquad\qquad (11.13)$$

for describing the sizes of the matrices involved in the multiplication. Since the first number in any of the above brackets describing matrix size refers to the number of rows, and the second number in the brackets refers to the number of columns, and because multiplication requires the number of columns in the first matrix to be equal to the number of rows in the second, the two 'inner' numbers in the bracket combination on the left-hand side of *11.13*, i.e. (2×2) (2×3), must be the same. Moreover, the dimensions of the product matrix are given by the two 'outer' numbers in this bracket combination, (2×3) in this example.

In general, for two matrices, **A** and **B**, of sizes $(m_A \times n_A)$ and $(m_B \times n_B)$, respectively, the product **AB** can be formed only if $n_A = m_B$, and the resulting product matrix is of size $(m_A \times n_B)$.

To demonstrate the multiplication rule for two general matrices, as was done for addition in *11.11*, would be very cumbersome, so we shall show the rule in symbol notation for the two small matrices shown below. Let

$$
\mathbf{A} = \begin{bmatrix} a_{11} & a_{12} & a_{13} \\ a_{21} & a_{22} & a_{23} \end{bmatrix}
\qquad
\mathbf{B} = \begin{bmatrix} b_{11} & b_{12} \\ b_{21} & b_{22} \\ b_{31} & b_{32} \end{bmatrix}
$$

Define a matrix **C** such that **AB** = **C**, then

$$
\mathbf{C} = \begin{bmatrix} (a_{11}b_{11} + a_{12}b_{21} + a_{13}b_{31}) & (a_{11}b_{12} + a_{12}b_{22} + a_{13}b_{32}) \\ (a_{21}b_{11} + a_{22}b_{21} + a_{23}b_{31}) & (a_{21}b_{12} + a_{22}b_{22} + a_{23}b_{32}) \end{bmatrix} \qquad (11.14)
$$

Notice that **A** is of size (2×3) and **B** is of size (3×2), thus $n_A = m_B = 3$, and the size of the product matrix is $(m_A \times n_B) = (2 \times 2)$.

Example 11.3

$$
\text{Define} \qquad \mathbf{A} = \begin{bmatrix} 4 & -6 \\ 2 & 5 \end{bmatrix}
\qquad
\mathbf{B} = \begin{bmatrix} 1 & 4 \\ 2 & -2 \\ 3 & -1 \end{bmatrix}
\qquad
\mathbf{c} = \begin{bmatrix} -4 \\ 6 \end{bmatrix}
$$

Find, if possible,
  (*a*) **AB**   (*b*) **BA**   (*c*) **Ac**
  (*d*) **cA**   (*e*) **Bc**   (*f*) **cB**

(*a*) **A** is a (2×2) matrix, and **B** is a (3×2) one; and the potential product can be represented as $(2 \times 2)(3 \times 2)$. So the number of columns in the first matrix is not equal to the number of rows in the second, and thus the product **AB** cannot be formed.

(*b*) This product is possible, and we have

$$
\begin{bmatrix} 1 & 4 \\ 2 & -2 \\ 3 & -1 \end{bmatrix}\begin{bmatrix} 4 & -6 \\ 2 & 5 \end{bmatrix} = \begin{bmatrix} (4+8) & (-6+20) \\ (8-4) & (-12-10) \\ (12-2) & (-18-5) \end{bmatrix} = \begin{bmatrix} 12 & 14 \\ 4 & -22 \\ 10 & -23 \end{bmatrix}
$$

(*c*) This product is also feasible:

$$
\begin{bmatrix} 4 & -6 \\ 2 & 5 \end{bmatrix}\begin{bmatrix} -4 \\ 6 \end{bmatrix} = \begin{bmatrix} (-16-36) \\ (-8+30) \end{bmatrix} = \begin{bmatrix} -52 \\ 22 \end{bmatrix}
$$

(*d*) This product cannot be formed.
(*e*) This product can be formed:

$$\begin{bmatrix} 1 & 4 \\ 2 & -2 \\ 3 & -1 \end{bmatrix} \begin{bmatrix} -4 \\ 6 \end{bmatrix} = \begin{bmatrix} (-4 + 24) \\ (-8 - 12) \\ (-12 - 6) \end{bmatrix} = \begin{bmatrix} 20 \\ -20 \\ -18 \end{bmatrix}$$

(*f*) This product is impossible.

When two matrices can be multiplied together in the order given, they are said to be conformable for multiplication. Thus, in *example 11.3*, **BA** is conformable, whereas **AB** is not conformable for multiplication.

Another point of terminology arises because the order in which the matrices are taken for multiplication is important. In *example 11.3*(b), **A** is said to be *pre-multiplied* by **B**, or one could equally say that **B** is *post-multiplied* by **A**.

Example 11.4
Let

$$A = \begin{bmatrix} -1 & 1 \\ 2 & -3 \end{bmatrix} \qquad B = \begin{bmatrix} 2 & 3 \\ -4 & 5 \end{bmatrix}$$

$$I = \begin{bmatrix} 1 & 0 \\ 0 & 1 \end{bmatrix} \qquad O = \begin{bmatrix} 0 & 0 \\ 0 & 0 \end{bmatrix}$$

Find

(*a*) **AB**   (*b*) **BA**   (*c*) **AI**
(*d*) **IA**   (*e*) **BO**   (*f*) **OB**

All these products are feasible: why?

(*a*)
$$\begin{bmatrix} -1 & 1 \\ 2 & -3 \end{bmatrix} \begin{bmatrix} 2 & 3 \\ -4 & 5 \end{bmatrix} = \begin{bmatrix} (-2 - 4) & (-3 + 5) \\ (4 + 12) & (6 - 15) \end{bmatrix} = \begin{bmatrix} -6 & 2 \\ 16 & -9 \end{bmatrix}$$

(*b*)
$$\begin{bmatrix} 2 & 3 \\ -4 & 5 \end{bmatrix} \begin{bmatrix} -1 & 1 \\ 2 & -3 \end{bmatrix} = \begin{bmatrix} (-2 + 6) & (2 - 9) \\ (4 + 10) & (-4 - 15) \end{bmatrix} = \begin{bmatrix} 4 & -7 \\ 14 & -19 \end{bmatrix}$$

(*c*)
$$\begin{bmatrix} -1 & 1 \\ 2 & -3 \end{bmatrix} \begin{bmatrix} 1 & 0 \\ 0 & 1 \end{bmatrix} = \begin{bmatrix} (-1 + 0) & (0 + 1) \\ (2 - 0) & (0 - 3) \end{bmatrix} = \begin{bmatrix} -1 & 1 \\ 2 & -3 \end{bmatrix}$$

(*d*)
$$\begin{bmatrix} 1 & 0 \\ 0 & 1 \end{bmatrix} \begin{bmatrix} -1 & 1 \\ 2 & -3 \end{bmatrix} = \begin{bmatrix} (-1 + 0) & (1 + 0) \\ (0 + 2) & (0 - 3) \end{bmatrix} = \begin{bmatrix} -1 & 1 \\ 2 & -3 \end{bmatrix}$$

(e)
$$\begin{bmatrix} 2 & 3 \\ -4 & 5 \end{bmatrix} \begin{bmatrix} 0 & 0 \\ 0 & 0 \end{bmatrix} = \begin{bmatrix} 0 & 0 \\ 0 & 0 \end{bmatrix}$$

(f)
$$\begin{bmatrix} 0 & 0 \\ 0 & 0 \end{bmatrix} \begin{bmatrix} 2 & 3 \\ -4 & 5 \end{bmatrix} = \begin{bmatrix} 0 & 0 \\ 0 & 0 \end{bmatrix}$$

In *example 11.4*: (a) and (b) show that even if two matrices can be multiplied both ways, the answer is different in each case. This emphasizes that, in general, matrix multiplication is not commutative.

In (c) and (d) it is shown that the unit matrix, I, behaves in matrix algebra in a similar manner to the number 1 in ordinary algebra. A matrix, A, can be pre- or post-multiplied by I, and the result will always by A; so that multiplication involving I *is* commutative,

i.e. $\quad AI = IA = A \quad$ (11.15)

A similar situation exists for the zero matrix in (e) and (f). This matrix behaves analogously to zero of ordinary algebra. Any matrix, A, pre- or post-multiplied by O yields the zero matrix; so that, again, multiplication involving O is commutative,

i.e. $\quad AO = OA = O \quad$ (11.16)

Multiplication of two vectors

The same rules apply to vectors as to matrices, and a worked example may help clarify further in the reader's mind the rules of matrix multiplication.

Example 11.5

Define $\quad a = \begin{bmatrix} 1 \\ -3 \\ 2 \end{bmatrix} \quad b = \begin{bmatrix} 4 \\ 0 \\ -1 \end{bmatrix}$

Find, if possible
    (a) ab (b) $a'b$ (c) ba'
    (d) ab' (e) $b'a$

(a) Vector a is a (3×1) matrix, so too is vector b; the potential product can be represented as $(3 \times 1)(3 \times 1)$. So the number of columns in the first matrix is not equal to the number of columns in the second, and thus the product ab cannot be formed.

(*b*) Here we have a product of $(1 \times 3) (3 \times 1)$ which is feasible, and it seems as though we shall obtain (1×1) matrix as answer, which is a scalar.

$$[1 \quad -3 \quad 2] \begin{bmatrix} 4 \\ 0 \\ -1 \end{bmatrix} = (1) (4) + (-3) (0) + (2) (-1) = 2$$

(*c*)

$$\begin{bmatrix} 4 \\ 0 \\ -1 \end{bmatrix} [1 \quad -3 \quad 2] = \begin{bmatrix} (4) (1) & (4) (-3) & (4) (2) \\ (0) (1) & (0) (-3) & (0) (2) \\ (-1) (1) & (-1) (-3) & (-1) (2) \end{bmatrix}$$

$$= \begin{bmatrix} 4 & -12 & 8 \\ 0 & 0 & 0 \\ -1 & 3 & -2 \end{bmatrix}$$

In this case, we obtain a (3×3) matrix as forecast by the product $(3 \times 1) (1 \times 3)$.

(*d*)

$$\begin{bmatrix} 1 \\ -3 \\ 2 \end{bmatrix} [4 \quad 0 \quad -1] = \begin{bmatrix} 4 & 0 & -1 \\ -12 & 0 & 3 \\ 8 & 0 & -2 \end{bmatrix}$$

We see that $\mathbf{ab'} = (\mathbf{ba'})'$, which is, in fact, a general rule.

(*e*)

$$[4 \quad 0 \quad -1] \begin{bmatrix} 1 \\ -3 \\ 2 \end{bmatrix} = 4 + 0 - 2 = 2$$

We see that $\mathbf{b'a} = \mathbf{a'b}$ which again is a general rule.

Multiplication of a matrix by its transpose

Example 11.6

Define $\mathbf{A} = \begin{bmatrix} 1 & 2 & 3 \\ 4 & -2 & -1 \end{bmatrix}$ and find (*a*) $\mathbf{AA'}$ and (*b*) $\mathbf{A'A}$

(a)
$$\mathbf{AA'} = \begin{bmatrix} 1 & 2 & 3 \\ 4 & -2 & -1 \end{bmatrix} \begin{bmatrix} 1 & 4 \\ 2 & -2 \\ 3 & -1 \end{bmatrix} = \begin{bmatrix} (1+4+9) & (4-4-3) \\ (4-4-3) & (16+4+1) \end{bmatrix}$$

$$= \begin{bmatrix} 14 & -3 \\ -3 & 21 \end{bmatrix}$$

(b)
$$\mathbf{A'A} = \begin{bmatrix} 1 & 4 \\ 2 & -2 \\ 3 & -1 \end{bmatrix} \begin{bmatrix} 1 & 2 & 3 \\ 4 & -2 & -1 \end{bmatrix} = \begin{bmatrix} (1+16) & (2-8) & (3-4) \\ (2-8) & (4+4) & (6+2) \\ (3-4) & (6+2) & (9+1) \end{bmatrix}$$

$$= \begin{bmatrix} 17 & -6 & -1 \\ -6 & 8 & 8 \\ -1 & 8 & 10 \end{bmatrix}$$

In general, a matrix may always be pre- or post-multiplied by its transpose and the result is a symmetric matrix. However, $\mathbf{AA'} = \mathbf{A'A}$ in general; the two product matrices are typically of different sizes. Further, product matrices of the form $\mathbf{AA'}$ or $\mathbf{A'A}$ are always symmetric.

Example 11.7

Define $\quad \mathbf{S} = \begin{bmatrix} 2 & 4 & -1 \\ 0 & 1 & 3 \\ -1 & -2 & 5 \end{bmatrix}$ and find (a) $\mathbf{SS'}$ and (b) $\mathbf{S'S}$

(a) $\quad \mathbf{SS'} = \begin{bmatrix} 2 & 4 & -1 \\ 0 & 1 & 3 \\ -1 & -2 & 5 \end{bmatrix} \begin{bmatrix} 2 & 0 & -1 \\ 4 & 1 & -2 \\ -1 & 3 & 5 \end{bmatrix}$

$$= \begin{bmatrix} (4+16+1) & (0+4-3) & (-2-8-5) \\ (0+4-3) & (0+1+9) & (0-2+15) \\ (-2-8-5) & (0-2+15) & (1+4+25) \end{bmatrix} = \begin{bmatrix} 21 & 1 & -15 \\ 1 & 10 & 13 \\ -15 & 13 & 30 \end{bmatrix}$$

(b) $$\mathbf{S'S} = \begin{bmatrix} 2 & 0 & -1 \\ 4 & 1 & -2 \\ -1 & 3 & 5 \end{bmatrix} \begin{bmatrix} 2 & 4 & -1 \\ 0 & 1 & 3 \\ -1 & -2 & 5 \end{bmatrix}$$

$$= \begin{bmatrix} (4+0+1) & (8+0+2) & (-2+0-5) \\ (8+0+2) & (16+1+4) & (-4+3-10) \\ (-2+0-5) & (-4+3-10) & (1+9+25) \end{bmatrix} = \begin{bmatrix} 5 & 10 & -7 \\ 10 & 21 & -11 \\ -7 & -11 & 35 \end{bmatrix}$$

This example shows that even when a square matrix is pre- and post-multiplied by its transpose, although the size of the product matrix is the same in each case, $\mathbf{SS'} \neq \mathbf{S'S}$ in general.

Inversion

Direct division of matrices is not defined, but under certain conditions, one matrix can be 'divided' into another by an indirect process involving multiplication. The concept is best demonstrated in relation to ordinary numbers.

Suppose we require $12 \div 4$. This could be evaluated as 12×0.25, i.e. $12 \times 1/4$ or 12×4^{-1}. The number 4^{-1} is the reciprocal, or *inverse*, of 4, and the expression $12 \times 4^{-1} = 12 \div 4 = 3$. The inverse of a number is such that when it is multiplied by the number itself, the product is unity; e.g. $4 \times 4^{-1} = 4 \times 1/4 = 1$. In general terms, we have

$$xx^{-1} = 1 \tag{11.17}$$

A square matrix can usually (but not always) be inverted, which means that when the inverse of the matrix is multiplied by the matrix itself then the product is the unit matrix, which we have already seen has similar properties in matrix algebra to the number 1 in ordinary algebra. So we have that

$$\mathbf{AA}^{-1} = \mathbf{I} \tag{11.18}$$

where $\mathbf{A}$ is a square matrix, and $\mathbf{A}^{-1}$ is the corresponding inverse matrix.

Inverse matrices are very important in matrix algebra, and this section has served to introduce the concept. The process for inverting a matrix is not straightforward, however, and consideration of this and other features of matrix algebra appears in *A Biologist's Advanced Mathematics*.

Some applications of matrix algebra

By far the most common applications of matrix algebra in biology are in the fields of population dynamics, and in the statistical analysis of multivariate populations. The latter is based on the theory dealt with in Chapter 3, and so relates to problems in taxonomy, genecology, and ecology. However, it is impossible to deal with the application of matrix algebra to the statistical analysis of multivariate populations here, as much intermediate statistical theory needs to be known. Instead, more straightforward applications will be presented.

Population growth with age-dependent birth and death rates

In considering the growth of a microbial population, we found that the logistic function could well describe the course of population growth with time. The relevant characteristic of the logistic function is that it allows for a slowing down of the growth rate with increasing density of the population, but the function does not permit other possible characteristics of population growth to be allowed for. In higher organisms, we have the situation that older individuals are more likely to die in a given interval of time than younger ones, and females of different ages may reproduce at different rates.

A model of population growth embodying these ideas can be formulated in terms of matrices. There are, however, two fundamental distinctions between this matrix model and the kind of model based on the logistic function. The latter regards time as a continuous variable, and so sets up a differential equation, whereas in the following matrix model time is thought of as a succession of discrete intervals (so that, of course, no differential equation can be formed). Thus the model is purely descriptive, and does not shed any light on the underlying dynamics of the population.

In what follows, the term 'individual' refers to any individual member of the population if the species is hermaphrodite, but only to a female member if the species is bisexual. In the latter case, however, the total numbers can be obtained if it is assumed that the ratio of the sexes, one to another, is constant, and that the death rates are the same for both sexes. At some time, t, a population can be represented by the column vector

$$\mathbf{n}_t = \begin{bmatrix} n_{0t} \\ n_{1t} \\ n_{2t} \\ \vdots \\ n_{mt} \end{bmatrix}$$

There are two subscripts to each element, and the second of these, t, is constant throughout and corresponds to the subscript of the symbol of the vector. The first subscript refers to the age of individuals in the population. Hence, at time t there are n_{0t} individuals between the ages of 0 and 1 units of time, n_{1t} individuals between the ages of 1 and 2, and so on. The sum of the elements gives the total number of individuals in the population. In general form, an element in the column vector $\mathbf{n}_t$ is n_{xt}, which denotes the number of individuals (at time t) between the ages of x and $x + 1$ units of time. There are $m + 1$ elements in the column vector, which assumes that no individual can survive longer than m units of time. We now enquire about the elements of the column vector $\mathbf{n}_{t+1}$, that is, the number of individuals of each age in the next time interval.

Denote by f_x the number of offspring (daughters) born to a female of age x that will survive into the next unit of time; the offspring will have age zero, and individuals already in existence at time t will move up to the next age group. The quantity f_x will, of course, be an average value. Next, denote by p_x the probability that an individual of age x will survive to the next time interval (probability is a number that can lie in the range $0 \leqslant p \leqslant 1$, where 0 denotes impossibility of an event, and 1 denotes certainty of an event). Write a matrix $\mathbf{P}$, as follows:

$$\mathbf{P} = \begin{bmatrix} f_0 & f_1 & f_2 \cdots f_{m-1} & f_m \\ p_0 & 0 & 0 \ldots \quad 0 & 0 \\ 0 & p_1 & 0 \ldots \quad 0 & 0 \\ \vdots & \vdots & \vdots \qquad \vdots & \vdots \\ 0 & 0 & 0 \ldots p_{m-1} & 0 \end{bmatrix}$$

Then the column vector, $\mathbf{n}_{t+1}$, describing the population at the next unit of time is given by

$$\mathbf{n}_{t+1} = \mathbf{P}\mathbf{n}_t \qquad (11.19)$$

for we have

$$\begin{bmatrix} f_0 & f_1 & f_2 \cdots f_{m-1} & f_m \\ p_0 & 0 & 0 \ldots \quad 0 & 0 \\ 0 & p_1 & 0 \ldots \quad 0 & 0 \\ \vdots & \vdots & \vdots \qquad \vdots & \vdots \\ 0 & 0 & 0 \ldots p_{m-1} & 0 \end{bmatrix} \begin{bmatrix} n_{0t} \\ n_{1t} \\ n_{2t} \\ \vdots \\ n_{mt} \end{bmatrix} = \begin{bmatrix} f_0 n_{0t} + f_1 n_{1t} + \cdots + f_m n_{mt} \\ p_0 n_{0t} \\ p_1 n_{1t} \\ \vdots \\ p_{m-1} n_{(m-1)t} \end{bmatrix}$$

and so

$$\mathbf{n}_{t+1} = \begin{bmatrix} n_{0(t+1)} \\ n_{1(t+1)} \\ n_{2(t+1)} \\ \vdots \\ n_{m(t+1)} \end{bmatrix} = \begin{bmatrix} f_0 n_{0t} + f_1 n_{1t} + \cdots + f_m n_{mt} \\ p_0 n_{0t} \\ p_1 n_{1t} \\ \vdots \\ p_{m-1} n_{(m-1)t} \end{bmatrix}$$

The matrix $\mathbf{P}$ may be called a transition matrix.

For calculating population structure after further time units, we first have

$$\mathbf{n}_{t+2} = \mathbf{P}\mathbf{n}_{t+1} \qquad (11.20)$$

Substituting for $\mathbf{n}_{t+1}$ on the right-hand side of *11.20*, using *11.19*, gives

$$\mathbf{n}_{t+1} = \mathbf{P}\mathbf{P}\mathbf{n}_t = \mathbf{P}^2\,\mathbf{n}_t$$

Hence, after s units of time

$$\mathbf{n}_{t+s} = \mathbf{P}^s\,\mathbf{n}_t \qquad (11.21)$$

It is also possible to work backwards from the time, t, when the population was first observed. Matrix $\mathbf{P}$ is square, so should be invertible, to $\mathbf{P}^{-1}$. Now pre-multiply both sides of *11.19* by $\mathbf{P}^{-1}$:

$$\mathbf{P}^{-1}\,\mathbf{n}_{t+1} = \mathbf{P}^{-1}\,\mathbf{P}\mathbf{n}_t$$

i.e. $\qquad \mathbf{n}_t = \mathbf{P}^{-1}\,\mathbf{n}_{t+1}$

So, commencing at time t: $\qquad \mathbf{n}_{t-1} = \mathbf{P}^{-1}\,\mathbf{n}_t \qquad (11.22)$

and $\qquad \mathbf{n}_{t-2} = \mathbf{P}^{-1}\,\mathbf{n}_{t-1}$

$$= \mathbf{P}^{-1}\,\mathbf{P}^{-1}\,\mathbf{n}_t$$

or $\qquad = (\mathbf{P}^{-1})^2\,\mathbf{n}_t$

Hence, after s previous units of time

$$\mathbf{n}_{t-s} = (\mathbf{P}^{-1})^s\,\mathbf{n}_t \qquad (11.23)$$

The values of the elements of the transition matrix are obtained by observation and experiment on the species concerned; or, they may be theoretical values, against which an actual population can be assessed by comparing the age structure of the population with column vectors produced by the theoretical model.

Species interaction in an ecosystem

Mathematically, this is a similar situation to the preceding one. Consider an ecosystem to have m co-existing species; at time t the number of each species present can be given by a column vector

$$\mathbf{n}_t = \begin{bmatrix} n_{1t} \\ n_{2t} \\ n_{3t} \\ \vdots \\ n_{mt} \end{bmatrix}$$

Define a transition matrix, $\mathbf{P}$, of order m so that the column vector of species numbers one unit of time later is given by

$$\mathbf{n}_{t+1} = \mathbf{P}\mathbf{n}_t$$

To interpret the meaning of $\mathbf{P}$, consider a specific, but highly artificial, example of an ecosystem with three species.

$$\text{Let} \qquad \mathbf{P} = \begin{bmatrix} 1 + \dfrac{t}{20} & 0 & 0 \\[3mm] 0 & 1 & \dfrac{t}{20} \\[3mm] 0 & 0 & 1 - \dfrac{t}{10} \end{bmatrix}$$

$$\text{Then} \qquad \mathbf{n}_{t+1} = \begin{bmatrix} 1 + \dfrac{t}{20} & 0 & 0 \\[3mm] 0 & 1 & \dfrac{t}{20} \\[3mm] 0 & 0 & 1 - \dfrac{t}{10} \end{bmatrix} \begin{bmatrix} n_{1t} \\[3mm] n_{2t} \\[3mm] n_{3t} \end{bmatrix} = \begin{bmatrix} n_{1t}\left(1 + \dfrac{t}{20}\right) \\[3mm] n_{2t} + \dfrac{n_{3t}\,t}{20} \\[3mm] n_{3t}\left(1 - \dfrac{t}{10}\right) \end{bmatrix}$$

It can be seen that species 1 increases in numbers independently of the other two. Species 3 decreases independently of the other two: whereas the rate of increase of species 2 depends on the current number of individuals of species 3. If $\mathbf{P}$ remains the same, eventually species 3 will become extinct and the number of individuals of species 2 will stabilize.

A biological interpretation of **P** can now be made. If **P** is a diagonal matrix, then each species changes independently of the others. It is the presence of off-diagonal elements that introduces species interaction. If **P** were a unit matrix, there would be no change in the species composition. In this example, species 2 changes in numbers only because species 3 does; species 2 is *partially* dependent upon species 3, since the former increases before the latter becomes extinct and then remains unchanged.

EXERCISES

1. If $\mathbf{P} = \begin{bmatrix} 1 & 2 & -1 \\ -3 & -2 & 1 \\ 2 & 3 & 2 \end{bmatrix}$ and $\mathbf{Q} = \begin{bmatrix} 3 & 2 & -1 \\ 4 & 2 & -2 \\ -1 & 3 & 1 \end{bmatrix}$

find (a) **P** + **Q**   (b) **Q** + **P**   (c) **P** − **Q**   (d) **Q** − **P**

2. If $\mathbf{A} = \begin{bmatrix} 2 & 3 \\ 1 & 6 \end{bmatrix}$ and $\mathbf{B} = \begin{bmatrix} 1 & 2 \\ 3 & 4 \end{bmatrix}$

find (a) **AB**   (b) **BA**

3. If $\mathbf{P} = \begin{bmatrix} -1 & 2 \\ 2 & -4 \end{bmatrix}$ and **A** and **B** are as in *example 2*

find (a) **(AB)P**   (b) **A(BP)**

4. If $\mathbf{A} = \begin{bmatrix} 3 & 2 & 1 \\ -2 & 3 & 6 \\ 1 & -2 & 4 \end{bmatrix}$ and $\mathbf{b} = \begin{bmatrix} 3 \\ 4 \\ 1 \end{bmatrix}$

find, if possible, (a) **Ab**   (b) **bA**

5. If **A** is as in *example 4* and $\mathbf{I} = \begin{bmatrix} 1 & 0 & 0 \\ 0 & 1 & 0 \\ 0 & 0 & 1 \end{bmatrix}$

find (a) **AI** (b) **IA**

6. If
$$\mathbf{C} = \begin{bmatrix} 4 & 0 & 2 & 1 \\ -1 & 6 & 3 & 8 \end{bmatrix} \qquad \mathbf{D} = \begin{bmatrix} 2 & 3 \\ 4 & 1 \\ 6 & 8 \\ 9 & 2 \end{bmatrix}$$

$$\mathbf{E} = \begin{bmatrix} 2 & 1 & 0 & 3 \\ 3 & 2 & 1 & 6 \\ -1 & 2 & 0 & 0 \\ 4 & 3 & 1 & 2 \end{bmatrix} \qquad \mathbf{F} = \begin{bmatrix} 2 & 1 & 2 \\ 3 & 0 & 6 \\ 2 & 1 & 8 \\ 4 & 1 & 6 \end{bmatrix}$$

find, if possible, (*a*) **CD**   (*b*) **CE**   (*c*)   **CF**
(*d*) **DE**   (*e*) **ED**   (*f*) **EF**

Answers to exercises

Chapter 2

1. 1, 2, 6, 24, 120, 720, 5040, 40 320, 362 880, 3 628 800
2. (a) 90 (b) 220
3. 0.2368
4. Put $\log_a b = x$ and $\log_b a = y$, then $b = a^x$ and $a = b^y$. Substituting the first of the two latter expressions into the second gives $a = a^{xy}$. Take logs of both sides: $\log a = xy \log a$, thus $xy = 1$, and the result is proved.

5. (a) Put $\log_a a = x$ then $a = a^x$. Thus $x = 1$.
   (b) Since $\log_a a^x = x \log_a a$, and $\log_a a = 1$, the result is proved.

Chapter 3

1. (a) 7.616 (b) 4.583 (c) 7.348
2. This is best performed as a Class Exercise, and the results may be written in the form of a two-way table:

	V. beccabunga	V. officinalis	V. montana	V. chamaedrys	V. serpyllifolia	V. hederifolia	V. persica
V. officinalis	53.0						
V. montana	34.1	28.8					
V. chamaedrys	43.7	13.4	18.8				
V. serpyllifolia	77.5	27.1	55.7	38.9			
V. hederifolia	70.8	20.4	48.3	31.8	9.4		
V. persica	67.4	15.6	41.0	25.0	18.5	12.7	
V. filiformis	69.7	18.2	42.5	26.8	19.1	15.9	7.5

3. (a) $y = -2.1146 + 1.7321x$ (b) $y = 0.6667 + 0.6667x$

Chapter 4

1. (a) 1   (b) 13   (c) −35
2. (a) 34   (b) 16
3. (a) 0.1250   (b) 0.2857 mg CO_2 cm^{-2} h^{-1}
4. $v = 0.062$ μmol min^{-1} in the absence of glucose; $v = 0.049$ μmol min^{-1} in the presence of glucose.

Chapter 5

2. (a) $9(3x - 2)^2$ (b) $-(2ax + b)/(ax^2 + bx + c)^2$
   (c) $n(1 - 1/x^2)(x + 1/x)^{(n-1)}$
3. (a) $4x/(1 - x^2)$ (b) $(2 - x)/\{2\sqrt{(1 - x)^3}\}$
   (c) $x(2a^2 - 3x^2)/\sqrt{(a^2 - x^2)}$
   (d) $\{n(ad - bc)/(cx + d)^2\}\{(ax + b)/(cx + d)\}^{(n-1)}$
4. (a) $-2/x^3, 6/x^4$ (b) $ab(1 - x)/(b + x)^2, ab(2x - 3)/(b + x)^3$

Chapter 6

1. For drawing the graphs, the following table provides some data in the range of n specified. For accurate graphs, more points will be required.

	$0.2n$	$1 - n/1\,000\,000$	$0.2n(1 - n/1\,000\,000)$	$1 - (n/1\,000\,000)^2$	$0.2n\{1 - (n/1\,000\,000)^2\}$
100 000	20 000	0.9	18 000	0.99	19 800
200 000	40 000	0.8	32 000	0.96	38 400
300 000	60 000	0.7	42 000	0.91	54 600
400 000	80 000	0.6	48 000	0.84	67 200
500 000	100 000	0.5	50 000	0.75	75 000
600 000	120 000	0.4	48 000	0.64	76 800
700 000	140 000	0.3	42 000	0.51	71 400
800 000	160 000	0.2	32 000	0.36	57 600
900 000	180 000	0.1	18 000	0.19	34 200

Combinations of *equations 6.20* and *6.21* gives

$$\frac{dn}{dt} = 0.2n\left(1 - \frac{n}{1\,000\,000}\right)$$

and the co-ordinates of the maximum point are $(500\,000,\ 50\,000)$.
Combination of *equations 6.20* and *6.22* gives

$$\frac{dn}{dt} = 0.2n\left\{1 - \left(\frac{n}{1\,000\,000}\right)^2\right\}$$

and the cordinates of the maximum point are $(577\,350,\ 76\,980)$

2. Maximum point: $(-1, 3)$ Minimum point: $(1, -1)$
   Point of inflexion: $(0, 1)$

Chapter 7

1. (a) $2x^3 + c$ (b) $-5/x + c$ (c) $x^4/4 - a^3 x + c$
   (d) $x^3/3 - 2x - 1/x + c$ (e) $\frac{2}{3}\sqrt{x^3} + \frac{4}{5}\sqrt{x^5} + \frac{2}{7}\sqrt{x^7} + c$
   (f) $(2x - 1)^6/12 + c$ (g) $-3\sqrt[3]{(1 - 4x)^2}/8 + c$
2. (a) $2\log_e (x - 2) + \log_e (x + 3) + c$
   (b) $\frac{1}{2}\log_e (x - 1) - \log_e (x - 2) + \frac{3}{2}\log_e (x - 3) + c$
3. (a) $16\frac{2}{3}$ (b) $14\frac{2}{3}$
4. (a) The answer will vary, depending on the accuracy of the graph and the counting technique, but should approximate to 0.5 square units.
   (b) Working to 8 decimal places 0.50041761
   (c) 0.50001247   (d) 0.5 exactly
5. (a) $8.\dot{3}$ (b) $2.9\dot{3}$

Chapter 8

1. (a) $y = a(1 - b\,e^{-x/a})$ where $b = e^{-c/a}$, and c is the constant of integration.
   (b) The integral involving x has to be split into partial fractions, and so the differential equation emerges in the form $dy/(y - 1) = dx/x - dx/(x + 1)$. The final solution is $y = \{x(c + 1) + 1\}/(x + 1)$. The constant of integration is $\log_e c$.
   (c) The integral involving y has to be split into partial fractions, yielding $dy/(1 - y) + dy/(1 + y) = dx/x$. The final solution is $(1 + y)/(1 - y) = cx$, $\log_e c$ being the constant of integration.
2. From the information given, we have $a + 2d = -11$ and $a + 6d = 5$. Solving this pair of simultaneous equations yields $a = -19$ and $d = 4$; thus the series is $-19, -15, -11, -7, -3, 1, 5, 9, \ldots$ The twentieth term is 57, and $S_{20} = 380$.
3. (a) $x_1^2 + x_2^2 + x_3^2 + x_4^2 + x_5^2$
   (b) $a(x^0 + x^1 + x^2 + x^3) = a(1 + x + x^2 + x^3)$
   (c) $\sqrt{x_1}/(a + x_1) + \sqrt{x_2}/(a + x_2) + \sqrt{x_3}/(a + x_3) + \sqrt{x_4}/(a + x_4)$
   (d) $x_1 x_2 x_3 x_4$

4. (a) $\frac{1}{3}\sum_{i=1}^{3} x_i$ (b) $\sum_{i=2}^{-2} ix^{(i+3)}$ (c) $\frac{1}{3}\prod_{i=1}^{3} x_i$ (d) $\sqrt[3]{\prod_{i=1}^{3} x_i}$

5. From the information given $a/(1 - r) = 36$ and $ar = 8$. These simultaneous equations give two solutions: either $a = 24$ and $r = 1/3$, or $a = 12$ and $r = 2/3$. So, two series fulfil the conditions:

$$24, 8, \tfrac{8}{3}, \tfrac{8}{9}, \tfrac{8}{27}, \ldots$$
or
$$12, 8, \tfrac{16}{3}, \tfrac{32}{9}, \tfrac{64}{27}, \ldots$$

6. (*a*) $\frac{1}{9}$ (*b*) $\frac{27}{99}$ (*c*) $\frac{123}{999}$
7. (*a*) $x^5 + 5x^4 + 10x^3 + 10x^2 + 5x + 1$
   (*b*) $x^5 - 5x^4 + 10x^3 - 10x^2 + 5x - 1$
   (*c*) $81x^4 - 216x^3 y + 216x^2 y^2 - 96xy^3 + 16y^4$
   (*d*) $x^8 + 12x^6 y^2 + 54x^4 y^4 + 108x^2 y^6 + 81y^6$

Chapter 9

1. (*a*) Birch: 0.3041   Sycamore: 0.2243 day^{-1}
   (*b*) Birch: 0.0441   Sycamore: 0.3837 g
   (*c*) Birch: 10.5   Sycamore: 21.8 g
   (*d*) 27.1 days
2. 14.2 days
3. (*a*) 13.9 (*b*) 12.0 mm
4. (*a*) $b\,e^{(a+bx)}$ (*b*) $2(x-1)\,e^{(1-2x+x^2)}$ (*c*) $-bk\,e^{-kx}$
   (*d*) $(b + 2cx)/(a + bx + cx^2)$
5. (*a*) $\log_e (1 + x) + c$ (*b*) $\frac{1}{3}\log_e (3x - 7) + c$ (*c*) $t - b\,e^{-kt}/k + c$
6. (*a*) 230 (*b*) 29 (*c*) 102 (*d*) 47

Chapter 10

t	L_A/L_W cm² g⁻¹ Unshaded	Shaded	L_W/W Unshaded	Shaded	$\bar{R}$ day⁻¹ Unshaded	Shaded	$\bar{E}$ g cm⁻² day⁻¹ Unshaded	Shaded
0	333.3	727.6	0.7500	0.6364	0.1703	0.1034	0.000681	0.000213
3	357.1	690.6	0.7000	0.7333	0.2189	0.0000	0.000776	0.000000
7	399.7	753.9	0.7708	0.7333	-0.0445	0.1834	-0.000200	0.000366
10	446.3	629.5	0.7381	0.7308	0.2067	0.1533	0.000729	0.000324
14	345.4	684.2	0.7292	0.7083	0.1465	0.0959	0.000612	0.000220
17	328.7	573.3	0.6980	0.6875	0.1758	-0.0215	0.000707	0.000037
21	389.0	684.4	0.6811	0.7167	-0.0839	0.1897	-0.000327	0.000392
24	370.7	684.6	0.6709	0.6981	0.1690	0.0678	-0.000714	0.000149
28	331.9	606.0	0.6848	0.7194				

Chapter 11

1. (a) and (b) $\begin{bmatrix} 4 & 4 & -2 \\ 1 & 0 & -1 \\ 1 & 6 & 3 \end{bmatrix}$ (c) $\begin{bmatrix} -2 & 0 & 0 \\ -7 & -4 & 3 \\ 3 & 0 & 1 \end{bmatrix}$

   (d) $\begin{bmatrix} 2 & 0 & 0 \\ 7 & 4 & -3 \\ -3 & 0 & -1 \end{bmatrix}$

2. (a) $\begin{bmatrix} 11 & 16 \\ 19 & 26 \end{bmatrix}$ (b) $\begin{bmatrix} 4 & 15 \\ 10 & 33 \end{bmatrix}$

3. (a) and (b) $\begin{bmatrix} 21 & -42 \\ 33 & -66 \end{bmatrix}$

4. (a) $\begin{bmatrix} 18 \\ 12 \\ -1 \end{bmatrix}$ (b) not conformable for multiplication

5. (a) and (b) $\begin{bmatrix} 2 & 2 & 1 \\ -2 & 3 & 6 \\ 1 & -2 & 4 \end{bmatrix}$

6. (a) $\begin{bmatrix} 29 & 30 \\ 112 & 43 \end{bmatrix}$ (b) $\begin{bmatrix} 10 & 11 & 1 & 14 \\ 45 & 41 & 14 & 49 \end{bmatrix}$ (c) $\begin{bmatrix} 16 & 7 & 30 \\ 54 & 10 & 106 \end{bmatrix}$

   (d) not conformable for multiplication.

   (e) $\begin{bmatrix} 35 & 13 \\ 74 & 31 \\ 6 & -1 \\ 44 & 27 \end{bmatrix}$ (f) $\begin{bmatrix} 19 & 5 & 28 \\ 38 & 10 & 62 \\ 4 & -1 & 10 \\ 27 & 7 & 46 \end{bmatrix}$

Index

Bold numbers indicate a main section, which may extend over more than one page, and/or a definition. Italicized numbers refer to pages containing relevant text figures. Although for most entries an attempt has been made to make the references comprehensive, this has not been done for entries with very numerous references and for which only important ones are given.